Modelling Order and Disorder

Modelling Order and Disorder: Integro–Differential Nonlinear Equations provides an overview of a general mathematical structure: integro-differential nonlinear equations. This mathematical structure provides a unified approach to model complex systems in social sciences, economics, biology, medicine, and other quantitative disciplines. The general aim of the book is to reflect possible organization and disorganization phenomena in the applied sciences, as well as to focus on non–local interactions.

Features

- Applications to social, biological, and physical phenomena
- Suitable for researchers and post-graduate students
- Open questions and perspectives on future avenues of research

Monographs and Research Notes in Mathematics
Series Editors: John A. Burns, Thomas J. Tucker, Miklos Bona, and Michael Ruzhansky

This series is designed to capture new developments and summarize what is known over the entire field of mathematics, both pure and applied. It will include a broad range of monographs and research notes on current and developing topics that will appeal to academics, graduate students, and practitioners. Interdisciplinary books appealing not only to the mathematical community, but also to engineers, physicists, and computer scientists are encouraged.

This series will maintain the highest editorial standards, publishing well-developed monographs as well as research notes on new topics that are final, but not yet refined into a formal monograph. The notes are meant to be a rapid means of publication for current material where the style of exposition reflects a developing topic.

Generalized Notions of Continued Fractions
Ergodicity and Number Theoretic Applications
Juan Fernández Sánchez, Jerónimo López-Salazar Codes, Juan B. Seoane Sepúlveda, and Wolfgang Trutschnig

Aspects of Integration
Novel Approaches to the Riemann and Lebesgue Integrals
Ronald B. Guenther and John W. Lee

Direct and Projective Limits of Geometric Banach Structures
Patrick Cabau and Fernand Pelletier

Classical Clifford Algebras
Operator-Algebraic and Free-Probabilistic Approaches
Ilwoo Cho

Perturbed Functional Iterations
A Matrix Free Large-Scale Nonlinear System Solver in Applied Science with an Introduction to D-Mapping
Suhrit Dey

Separate and Joint Continuity
Jiling Cao and Warren B. Moors

Fractional Integrals, Potentials, and Radon Transforms, Second Edition
Boris Rubin

Variational-Hemivariational Inequalities with Applications, Second Edition
Mircea Sofonea and Stanisław Migórski

Classification of Lipschitz Mappings, Second Edition
Torrey M. Gallagher, Víctor Pérez-García, and Łukasz Piasecki

Completely Regular Codes in Distance Regular Graphs
Patrick Solé and Minjia Shi

Non-Newtonian Sequence Spaces with Applications
Feyzi Başar and Bipan Hazarika

Spectral Theory for Linear Operators
Bilel Krichen

Convex Analysis in Polynomial Spaces with Applications
Domingo García, Mingu Jung, Manuel Maestre, Gustavo A. Muñoz Fernández and Juan B. Seoane Sepúlveda

Modelling Order and Disorder
Integro–Differential Nonlinear Equations
Mirosław Lachowicz

For more information about this series please visit: www.routledge.com/Chapman--HallCRC-Monographs-and-Research-Notes-in-Mathematics/book-series/CRCMONRESNOT

Modelling Order and Disorder

Integro–Differential Nonlinear Equations

Mirosław Lachowicz

CRC Press
Taylor & Francis Group
Boca Raton London New York

CRC Press is an imprint of the
Taylor & Francis Group, an **informa** business
A CHAPMAN & HALL BOOK

First edition published 2026
by CRC Press
2385 NW Executive Center Drive, Suite 320, Boca Raton FL 33431

and by CRC Press
4 Park Square, Milton Park, Abingdon, Oxon, OX14 4RN

CRC Press is an imprint of Taylor & Francis Group, LLC

ISBN: 978-1-032-88400-4 (hbk)
ISBN: 978-1-032-88402-8 (pbk)
ISBN: 978-1-003-53759-5 (ebk)

DOI: 10.1201/9781003537595

Typeset in CMR10
by KnowledgeWorks Global Ltd.

Publisher's note: This book has been prepared from camera-ready copy provided by the authors.

Dedication

Dedykuję Mojej Kasi

***

Italian proverb: Dove non c'é ordine, c'é disordine.

Commemoration

In the process of writing this monograph, I recall the memory of my friend and collaborator Henryk Leszczyński (1960–2021). We struggled together with many of the results included in this monograph. We represented different approaches, but we complemented each other. Additionally, I would like to mention colleagues with whom I did not write joint papers but whom I was lucky to meet on my scientific path: Konstanty Holly (1954–1998), Tadeusz Nadzieja (1951–2021), Wojciech Okrasiński (1950–2020), Kazimierz Piechór (1943–2016) and Kazimierz Sobczyk (1939–2017).

Contents

Preface

This monograph provides an overview of a general mathematical structure, integro–differential nonlinear equations, which can be used to model both order and disorder phenomena in the life and social sciences. The mathematical structures exhibit intriguing mathematical behaviour. The possibilities of blow-ups in a finite time of a solution can be interpreted as alignment and refer to the creation of order (self–organization), whereas the existence of global in–time solutions can lead to the description of kind of disorder systems. The general mathematical structure can be applicable to various particular processes. In the simplest case, the constant rate of interaction of agents is fully characterized. In fact, we are able to give a full mathematical description of the behaviour of solutions depending on the type of initial data. The more relevant case, the small range of interaction, is considered as well. However, it is a more complex situation that can lead to an interesting result that may be of significant importance in the description of life and social phenomena. The asymptotic relationships with partial differential equations, which are also a typical way of approaching reality, are discussed as well. We present microscopic models that correspond to the previously defined mesoscopic models. We discuss the diauxic behaviour at various scales. We also briefly discuss the discrete models. An intriguing application of the mathematical frameworks presented in the monograph is their potential to model spatially non–local effects. In particular, non–local models of cell movement related to cancer invasion into surrounding tissue are discussed in detail. Nevertheless, this methodology is sufficiently broad to be adapted for characterizing the movement of various biological entities where non–local effects are significant. Finally, we present some particular examples of processes that can be modelled by structures considered in the monograph. In particular, we consider the competition between the tumour and the immune system, the tendon healing process called collagen remodelling that takes place in the last stage of tendon recovery, the DNA denaturation process, and the relocation of individuals in a closed domain such as an elevator. These particular examples show how the structures considered in the monograph can be used to describe processes in the life and social sciences.

The equations studied in the monograph lead to phenomena of order or disorder, depending on the situation considered. We discuss whether and how it is possible to mathematically model order, that is, self–organization, or disorder, that is, disorganization, in life and social systems. The presented theory of integro–differential equations, the so–called kinetic equations, allows us to provide a mathematical basis for such behaviour.

Author

Mirosław A. Lachowicz is a Full Professor of Mathematics at the University of Warsaw. During 2016–2024 he was the chairman of the Scientific Committee of the Institute of Applied Mathematics and Mechanics; 2019–2023 a member of the Mathematical Committee of the Polish Academy of Sciences; and 2005–2016 director of the Institute of Applied Mathematics and Mechanics.

He is a member of the editorial boards of several institutions including: Mathematical Methods in the Applied Sciences, Mathematical Biosciences and Engineering, Letters in Biomathematics, and Opuscula Mathematica.

His research Interests include Integro-differential Equations, Singularly Perturbed Problems, Markov Processes, Applications in Life and Social Sciences, and relationships between Mathematics and the Humanities.

He had numerous scientific contacts, especially with universities in Italy and South Africa. He was an honorary professor at the University of KwaZulu–Natal in Durban, and he also spent some time at the Universities of Pretoria and Stellenbosch. He has travelled to numerous institutions in Europe and Japan.

Acknowledgements

The author was partially supported by the National Science Centre, Poland; Grant 2017/25/B/ST1/00051 and partially by Nowe Idee 3 B, 01/IDUB/–2019/94, University of Warsaw.

Chapter 1

Introduction

In certain fields, such as Social Sciences, Economics, Biology and Medicine, it is essential to take into account the interactions between the individual components (agents) of the specified systems. Mathematical structures referring to fully macroscopic statements could not be adequate. In such a situation, it is important to take into account the individual, microscopic states of the agents that compose the system in question. The proper mathematical structure should model the evolution of the states and give information on the behaviour of the whole system. The kinetic theory of gases suggests two potential approaches to consider. The first one, which may be referred to as purely microscopic, can consist of the description of the states of all agents in a stochastic way. It usually leads to the Markov processes and the mathematical structures defined by the evolution equations through the corresponding generators, the linear operators particularizing changes of the states due to the interactions or external influence. The second, relating directly to the idea of L. Boltzmann in kinetic theory, is based on a statistical description of the state of the test agent of the system. The state is given in terms of the probability density or the non–necessarily normalized distribution function in the classical Boltzmann setting. This approach can be termed mesoscopic, although it is sometimes thought of as microscopic. Formally, it can only be justified if the number of interacting agents is sufficiently large. However, following various approaches in physics, it can also be acceptable if the number of agents is moderately large. In such cases, any macroscopic strategy seems even less acceptable but is commonly used.

We may consider these two approaches, microscopic and mesoscopic, as alternatives and study the possible mathematical relationships between them; see Chapter 9 and Refs. [24, 129, 130] and the references therein. In many cases, however, they may lead to different results, and the proper choice of the model is, as always, a sophisticated art.

Here, we provide a unified approach to model complex systems in social sciences, economics, biology, medicine, and other disciplines within the framework of general mathematical structures, that is, integro–differential equations. The general aim is to reflect possible organization and disorganization phenomena in the Applied Sciences, as well as to focus on non–local interactions.

In thermodynamics, the second law states, through the notion of entropy, that physical processes move towards more disorder. On the other hand, biological phenomena, like life itself, can be related to the creation of order. Both these trends should be described by suitable mathematical models, i.e. the mathematical structure—usually equations—that are able to take into account both possible tendencies.

In the present book, we are going to provide a general mathematical framework and relate it with some important applications. The author's efforts should be considered preliminary. It seems that they may constitute an introduction to further research.

We consider an individual characteristic (state) of any agent in a large system. We want to define the distribution of the state through probability density. Such an approach is quite general and may be related to various particular applications. The main feature that is taken into account in the modelling is the interaction between agents of the system that can cause a change of state. This approach is generally referred to as the mesoscopic (kinetic) scale

DOI: 10.1201/9781003537595-1

of description. The general idea comes from Boltzmann's kinetic theory of gases, in which the mathematical structure defines the statistical behaviour of a test agent of the system in terms of the microscopic parameters v—the state of each agent of some population. As we already mentioned, the alternative approach can refer to the purely microscopic scale and the mathematical structure that takes into account the probabilistic description of the states of all agents in the system; cf. Chapter 9 and Ref. [24].

Structures of the type that we are going to discuss here were proposed by H. Othmer et al. in Ref. [164] and studied, e.g. in [96, 115]; see also references therein. For a bibliographical review, the reader is referred to [137, 170]. In Ref. [24, 129, 130], a general class of kinetic equations was considered.

In various applications, the microscopic parameter v can be related to particular individual characteristics of the agents; see Chapter 13. In Ref. [7] it is the general social state of the agent, whereas in Refs. [18], it is the pair—the position and velocity of a walker in human crowds. Ref. [34] deals with the dynamics of human crowds, the individual emotional state is taken into account when evacuating a metro station. The framework of kinetic theory in various situations can be found; see, e.g. [7, 31, 32, 34, 130, 139] and references therein. The general references to the mathematical approach of collective phenomena and self–organization in systems composed of large numbers of individuals are Refs. [9, 61, 62, 182]. An interesting application could be the vaccination issues that take into account the psychological individual aspects; see Refs. [154, 213].

In Ref. [23] (see also [24]), the agents are characterized by their position and orientation (+/−) and the corresponding macroscopic limit is studied. Various aspects of swarm behaviour are studied in Refs. [17, 21, 38, 42, 48, 49, 51, 68, 115, 136, 137, 155, 170]. The framework can also be associated with opinion formation—see Refs. [126, 138, 139, 169]. References [126, 192] deal with opinion dynamics with two types of responses to social influence: conformity and anti–conformity with the aim of illustrating polarized debates in society; see also references therein. A vehicular traffic can be described by the state of each vehicle and the interactions between vehicles on the individual or microscopic scale, at the level of kinetic description by the state of a sample of the traffic stream, the mesoscopic scale, and at the level of aggregate flow variables by the state of density, average velocity and flow, that is, the macroscopic scale; see, for example, [9, 61, 70, 71, 182].

An intriguing example of modelling is related to the spectacular organization of flocks of birds; see Refs. [17, 21, 38, 42, 155], where the states of agents (birds) are defined by their positions and velocities. This could also be an example of the need to take into account non–local types of interaction between agents; see Section 12, which seems typical for various biological and ecological phenomena; see Section 12. Another example is the movement of cells; see Section 12.3, for example, cancer cells.

A notable application of the mathematical frameworks outlined in the monograph is their potential for modelling spatial non–local effects—see Section 12. In this section such structures are used to describe the non–local effects of cell movement related to the processes of cancer invasion of the surrounding tissue. However, the methods presented are general enough that they can be used in other processes of movement of biological or social objects in which the effects of non–local interactions may be important. Non–local effects may appear in the case of complex geometric forms in the space of interacting objects. The interested reader can refer to Refs. [1, 5, 12, 52, 55, 81, 82, 85, 86, 117, 145, 166, 167, 171, 193] and references therein. In Qiang Du's book [81], dedicated to continuum mechanics, the author states: "Non–local models can account for non–local interaction and take on more general forms than discrete and local continuum models. They can be effective alternatives or serve as tools for understanding existing ones and bridges to help make connections". The non–local nature of quantum mechanics is widely discussed; see e.g. Refs. [147, 190, 198] and references therein. The papers by C.T. Lee et al., Ref. [150] and by J. Furter and M. Grinfeld Ref. [94], deal with the concept of non–local modelling in biology.

The non–local models, on the one hand, can lead to a more efficient mathematical description, for example, without the non–physical blow-ups of solutions caused by differential treatment in local models. On the other hand, non–local models can be treated as more general than local models; that is, they can be applied to a larger class of natural and social phenomena depending on the size of parameters characterizing the rates of spatial interactions and the relationships between them. Appropriate, experimentally justified, limit transitions of parameters determining the rates of interactions may lead to different local models, that is, in terms of partial differential equations. This seems to be a very interesting direction for further mathematical research with the potential for better descriptions of reality.

In Section 13.1 we consider a simple model of competition between the tumour and the immune system. Following Ref. [83], in Section 13.2, we show how the mathematical framework developed in Chapter 4 can be applied to model a very important tendon healing process. The model refers to the collagen remodelling that occurs in the latter stage of tendon recovery. It is the first model to study the process and can be a basis for further developed models that suggest a possible optimal surgery intervention.

In Ref. [66], see Section 13.3, a microscopic model related to the thermal denaturation process of DNA, or melting, is studied. The process refers to the splitting of DNA base pairs, resulting in the separation of two complementary DNA strands. The states of all base pairs of DNA, defined by microscopic parameters, characterize the distances between the bases within individual base pairs. Numerical simulations showed the sigmoid shape of the DNA melting curves and revealed the appearance of experimentally observed bubbles.

Microscopic and mesoscopic stochastic models of redistribution of individuals inside a domain which can be thought of as representing an elevator are proposed in Ref. [76]. The parameter characterizing an individual is its position in the domain. The microscopic approach takes into account the positions of all individuals, whereas the mesoscopic approach refers to a test individual. We discuss this model in Section 13.4.

Models taking into account symmetric and asymmetric interactions between agents in the context of opinion formation as well as the economic state and their limits leading to partial differential equations are studied in Refs. [138] and [134], respectively.

Generally, there are many possibilities of the applicability of this type of model, in which an individual parameter can have many interpretations, in addition to those mentioned above, in biology, e.g. dominance of agents, medicine, e.g. cell activity, social sciences, e.g. socio–economic status; see e.g. [7, 24, 31, 48, 75, 130, 146, 169, 170]—and references therein.

The book is organized as follows. Chapter 2 presents general models (Section 2.2) including those in terms of the formation of opinions (Section 2.3). In Section 2.4 the models with space structure are presented. Chapter 3 presents some preliminary results including the existence results. In Chapter 4 we discuss the simplest case, the rate of interaction (kernel) a that is constant. In such a case, we are able to give a full mathematical description of the behaviour of solutions depending on the type of initial data. Complicated dynamics are observed. The possibility of blow–ups can be directly related to the creation of swarms. The possible blow–ups versus the global existence of solutions are designated. Moreover, results on asymptotic behaviour are formulated. This type of analysis continues in Chapter 5, which examines the situation of the general kernels. Under suitable assumptions relating the kernel with initial data, blow–ups versus global existence, the results are formulated. Chapter 6 deals with the small range of interaction. However, it is a more complex situation that can lead to interesting results that may be of significant importance in the description of life and social phenomena. In particular, we show the possibility of creating a bimodal distribution from a unimodal distribution. In Chapter 7 we define the asymptotic relationships with the partial differential equations that can also be a basis for further modelling, as the partial differential equations are a typical way of approaching reality. Symmetric kernels (Section 7.1) and asymmetric kernels (Section 7.2) lead to different asymptotic results. Comments on equilibrium solutions are presented in Chapter 8. Chapter 9 presents microscopic models that correspond to the previously defined mesoscopic models. Chapter 10 is devoted to the diauxic behaviour on various scales. The discrete models are discussed in Chapter 11. Unlike the rest of the monograph, where the continuous case is considered, in Chapter 11 the case of a discrete and finite number of possible states is discussed. An interesting application of the mathematical frameworks presented in the monograph is their potential to model spatially non–local effects. In particular, non–local models of cell movement related to cancer invasion into surrounding tissue are discussed in detail in Chapter 12. Although we focus on cancer invasion, the methodology is sufficiently broad to be adapted to characterize the movement of various biological entities where non–local effects are significant; see Section 12.3. Chapter 13 presents some particular examples of processes that can be modelled by the mathematical structures considered in the monograph. In Section 13.1 a simple kinetic model of the competition between the tumour and the immune system is considered. Section 13.2 deals with models of the tendon healing process, also called collagen remodelling, that takes place in the last stage of tendon recovery. Section 13.3 refers to the DNA denaturation process. Section 13.4 deals with the relocation of individuals in a closed domain such as an elevator. These particular examples show how the structures considered in the monograph can be used to describe processes in the life and social sciences. The last chapter is devoted to concluding remarks.

In summary, the monograph describes the basic mathematical properties of integro-differential equations which describe the distribution of the individual state of an agent of a certain population. A change in state occurs through interactions with other agents. The equations studied are non–linear and can generate very rich and diverse behaviours of solutions with significant interpretations. The microscopic parameter that characterizes interacting agents can be referred, for example, to opinion or socio–economic state, or welfare state in social sciences; individual dominance, or separation between complementary DNA bases in the DNA denaturation in biology; cell activity in the process of cancer growth, or collagen fibre orientation in the process of tendon healing in medicine; individual emotional state, or learning state in psychology; agent's position, and/or velocity in vehicle traffic, or pedestrian dynamics; in the description of swarms, flocks, and crowds. The equations studied lead to phenomena of order or disorder, depending on the situation considered. The monograph discusses whether and how it is possible to mathematically model order, that is, self–organization, or disorder, that is, disorganization, in life and social systems. The presented theory of integro–differential equations, the so–called kinetic equations, allows us to provide a mathematical basis for such behaviour. The blow–ups of solutions can describe some self–organization effects in a finite time that can have a positive connotation, like healing, or a negative, like society polarization. The results can be applied to processes in social sciences, economics, biology; medicine. Moreover, the importance of space–non–local modelling of some biological phenomena including the invasion of cancer on the surrounding tissue is discussed as well. Various non–local models are presented.

Chapter 2

Basic mathematical structures

2.1 Kinetic equations

The equations considered in the monograph are often called kinetic because of their connection with their general idea derived from kinetic theory. The kinetic theory of rarefied gases was developed by Ludwig Boltzmann (1844–1906); see references in Refs. [54,72,116]. It refers to the description of a system of interacting particles on a mesoscopic scale. A simple gas, as a collection of interacting particles, can be described on a microscopic scale as the dynamics of all the particles involved. Therefore, it is necessary to provide the mechanical states of all particles, that is, their positions] and velocities, with a description of their interactions. Alternatively, on a macroscopic scale, the description is within continuum mechanics. The former is too detailed in specific situations and, moreover, requires experimental data that are practically impossible to obtain. The latter encapsulates aggregated data concerning a system made up of numerous agents and in many applications may be too simplified and not give good results. An example could be the evaporation phenomenon in which the Knudsen layer occurs, which goes beyond the scope of the macroscopic scale. An intermediate description is kinetic on a mesoscopic scale—indexmesoscopic scale in which the state of the test agent can be taken into account through a distribution function; see references in [32, 37, 54, 175]. In the case of classical kinetic theory, it refers to the mechanical state of the particle, and in other applications, it can refer to the activity, dominance, fitness, or other factors that characterize the agent. There is a vast literature on various versions of kinetic equations in life and social sciences; see, e.g. Refs. [13, 22, 28, 32, 33, 37, 44, 49, 53, 59, 175, 205]. Current advancements in modelling, qualitative analysis, and simulation of active particles, particularly focused on applications in life and social sciences, are detailed, for instance, in Refs. [31, 33, 53, 56, 103]. Collective behaviour of interacting agents in mathematical biology based on the evolution equations of observables was developed in Refs. [97, 98].

Most often, the mathematical structure of the kinetic equations used in life and social sciences is completely different from the mathematical structure of the Boltzmann equation. In particular, the mathematical framework of the equations studied in this monograph is unrelated to that of the Boltzmann equation. As a result, the mathematical problems considered in this monograph are of a completely different nature from those considered in classical kinetic theory.

An essential aspect of kinetic equations is their capacity to describe interactions among individual agents of the system (population). The interactions are crucial in both life and social sciences; therefore, it is of great importance to accurately characterize these interactions and to derive predictions regarding the behaviour of the system based on them.

Kinetic equations typically have a conservative (probabilistic) nature, which means that the integral of the integral interaction operator gives a contribution of 0. This means that the process does not take into account the birth or death of agents. Consequently, the natural option is the probabilistic characterization of the description. The situation in which the

DOI: 10.1201/9781003537595-2

proliferation or mortality of agents are taken into account can be reduced to a conservative (probabilistic) situation by introducing an additional auxiliary sub–population, "Hades", as practices in the theory of stochastic processes; see Example c in Section 8.1 in Ref. [215]. This auxiliary sub–population has no direct interpretation, but is a mental creation that allows the model to be conservative (probabilistic); cf. Ref. [47]. In a sense, this leads to an assumption analogous to the classic and widely accepted logistic model with a limited increase, a priori; see Section 12.4.

Kinetic equations can describe both organization that can be understood as order and disorganization that is viewed as the disorder. The discussion of these aspects, at the mathematical level, is the main goal of this monograph. We focus on a certain class of kinetic equations, integro–differential equations, which are introduced in this chapter; see Sections 2.2, 2.4 and 2.3. The class of equations considered here is quite specific; nevertheless, it plays a crucial role in characterizing all processes where the interactions among agents are a fundamental component.

It should be emphasized that various aspects of organization for other mathematical structures are abundantly represented in the literature; see, e.g. Refs. [9, 32, 49, 62, 73, 75, 105, 165] and references therein.

The book by N. Bellomo et al. [31] refers to particle methods to describe the dynamics of complex systems in economics. B. Perthame's book [175] deals with the kinetic–transport equation, showing interesting functional analytic properties and describing how macroscopic models are derived from kinetic equations. The paper Ref. [103] focuses on two particular dynamics of interaction closely related to the one described in the seminal paper by T. Vicsek et al. Ref. [208]. In Ref. [112] the hydrodynamic equations are derived from the Vicsek model by means of an Enskog–type kinetic theory. Y. Kozitsky with collaborators in a series of papers; see Refs. [123–125], studies an infinite system of particles, their stochastic properties, and various applications. Interesting applications of kinetic equations have been used to describe autoimmune diseases. Reference [60] introduced a novel mathematical model to evaluate the effects of drug therapies on autoimmune diseases. An optimal control problem was analysed and numerical tests were performed to explore strategies for optimal treatment. A mathematical model of kinetic type was proposed in Ref. [69, 156] to characterize the typical chronic aspect of autoimmune diseases. This model concerns the cellular interactions of the immune system within the autoimmune context and shows recurring dynamics. The study demonstrates that steady oscillations can emerge through a Hopf bifurcation. Reference [162] deals with the equilibrium states of a nonlinear kinetic system describing the relapsing–remitting stages observed in many chronic autoimmune diseases. In Ref. [163], see also Ref. [40, 41], a mathematical study of multiple sclerosis using a spatio–temporal kinetic theory model considering interactions among different populations of human cells and the motion of immune cells stimulated by cytokines. In a formal limit, macroscopic reaction–diffusion equations with chemotaxis terms are obtained. Spatial patterns are then studied through a Turing instability analysis. A model of practical application for rubber asphalt at the mesoscopic scale was constructed in Ref. [151].

2.2 Models

In this monograph, we consider a general nonlinear equation at the kinetic, mesoscopic level, that can reflect the time evolution of the probability density of the inner state variable v related to an individual test agent of the given system. The parameter is defined in the set

$\mathbb{V}$, where $\mathbb{V}$ is a domain in $\mathbb{R}^n$. We consider $(\mathbb{V}, \mu)$—a space with a σ–finite measure μ. The meaning of the function $f(t, v)$ is such that it is a probability density to find an agent at instant $t > 0$ with state $v \in \mathbb{V}$.

As we already mentioned, the microscopic parameter $v \in \mathbb{V}$ that characterizes interacting agents can be referred, for example, to opinion, cf. Chapter 2.3, or socio–economic state, or welfare state in social sciences; individual dominance, or separation between complementary DNA bases in the DNA denaturation in biology, cf. Section 13.3; cell activity in the process of cancer growth, cf. Section 13.1, or collagen fibre orientation in the process of tendon healing in medicine, cf. Section 13.2; individual emotional state, or learning state in psychology; agent's position, and/or velocity in vehicle traffic, or pedestrian dynamics; in the description of swarms, flocks, and crowds.

We start with the following general class of mesoscopic equations:

$$\frac{\partial}{\partial t} f(t, v) = \int_{\mathbb{V}} T_f(t, v, w) f(t, w) \, \mathrm{d}\mu(w) - f(t, v) \int_{\mathbb{V}} T_f(t, w, v) \, \mathrm{d}\mu(w) \,, \tag{2.1}$$

where $T_f(v, w)$ is the turning rate, which can depend on probability density f, of state $w \in \mathbb{V}$ to state $v \in \mathbb{V}$ resulting from the interactions with other agents. Therefore, T_f also depends on the variable t. To simplify the notation, in the following we will sometimes ignore this dependence in the notation. The modelling process requires the selection of an appropriate T_f. We employ the general σ–finite measure μ to cover three cases that can be of interest in applications,

[C] $\mathbb{V}$ is a nonempty Lebesgue measurable set and μ is the Lebesgue measure,

[D] $\mathbb{V}$ is a discrete set that is not empty and μ is the counting measure,

[CD] $\mathbb{V}$ is a Cartesian product of a discrete nonempty set $\mathbb{J}$ and a nonempty Lebesgue–measurable set $\mathbb{U}$ and μ is a product of the counting measure and the Lebesgue measure.

Equation (2.1) was initially suggested in Ref. [170]. Subsequent research in [136–140] explored various versions of it. An interesting alternative to Case **[C]** is the discrete characteristic variables in Case **[D]**; see Chapter 11 and cf. Ref. [141]. Finally, Case **[CD]** refers to the inner variable of the set $\mathbb{U}$, while the discrete component of the set $\mathbb{J}$ defines the sub–population of the agent—cf. [24, 130].

The natural choice (cf. [170] and [136–140]) could be

$$T_f(v, w) = \int_{\mathbb{V}} A(v, w, w_*) a(w, w_*) f^{\gamma}(t, w_*) \, \mathrm{d}\mu(w_*) \,,$$

where γ indicates the magnitude of the effect that other agents have influence on a given test agent. The higher the value of γ, the more powerful is the effect of influence. The function $a(w, w_*)$ is the interaction rate. We will also call it the interaction kernel. It describes the chance of interaction between agents with state w and agents with state w_*. The function $A(v, w, w_*)$ is the probability of transition (jump) from state w to state v due to interaction with the agent with state w_*.

We assume that both $a : \mathbb{V}^2 \to \mathbb{R}_+$ and $A : \mathbb{V}^3 \to \mathbb{R}_+$ are measurable functions and

$$A \geq 0 \,, \qquad \int_{\mathbb{V}} A(v, w, w_*) \, \mathrm{d}\mu(v) = 1 \,, \qquad \forall \, w, w_* \in \mathbb{V} \,. \tag{2.2}$$

Therefore, from Eq. (2.1) it follows

$$\frac{\partial}{\partial t} f(t,v) = \int_{\mathbb{V}}\int_{\mathbb{V}} A(v,w,w_*) a(w,w_*)\, f^\gamma(t,w_*)\, f(t,w)\, d\mu(w_*)\, d\mu(w)$$
$$-f(t,v) \int_{\mathbb{V}} a(v,w_*)\, f^\gamma(t,w_*)\, d\mu(w_*)\,. \tag{2.3}$$

Equation (2.3) can be mathematically connected to the behaviour of N random agents in the limit of N approaching infinity, as demonstrated in [24, 130], cf. Chapter 9. This relationship is established when the interactions between γ agents are taken into account, with $\gamma > 1$.

Referring to the example of the organization of flocks of birds, cf. [17, 21, 155], it seems that an agent (a bird) can adjust its state to the states of several neighbour birds (probably 6 or 7). In this context, the concept of topological interactions was developed; see [38, 42]. Such kind of interactions can lead in the modelling presented here to $\gamma = 7$.

We can assume

$$A(v,w,w') = \delta(v-w')\,,$$

and formally obtain

$$T_f(v,w) = a(w,v) f^\gamma(t,v)\,.$$

It leads to a completely different mathematical structure on the mesoscopic scale

$$\frac{\partial}{\partial t} f(t,v) = Q[f](t,v)\,. \tag{2.4}$$

where

$$Q[f](t,v) = f^\gamma(t,v) \int_{\mathbb{V}} a(w,v)\, f(t,w)\, d\mu(w)$$
$$-f(t,v) \int_{\mathbb{V}} a(v,w)\, f^\gamma(t,w)\, d\mu(w)\,. \tag{2.5}$$

Equation (2.4) in Case **[C]** was studied in Refs. [136–140, 170]. The simplest choice of the parameter γ is $\gamma = 1$, however, case $\gamma > 1$ leads to more interesting mathematical considerations that involve possibilities of blow–ups of solutions in a finite time, see [136–138]. The case $0 < \gamma < 1$ was considered in Ref. [170]. The function a, as already mentioned, is referred to as the interaction rate or the interaction kernel. Here, a more natural interpretation is that $a(v,w)$ denotes the rate of transition from state v to state w.

Introducing the notation

$$a \circledast_1 f(v) = \int_{\mathbb{V}} a(v,w)\, f(t,w)\, d\mu(w)\,,$$

and

$$a \circledast_2 f(v) = \int_{\mathbb{V}} a(w,v)\, f(t,w)\, d\mu(w)\,,$$

we can rewrite the nonlinear operator Eq. (2.5) in a compact form

$$Q[f] = f^\gamma\, a \circledast_2 f - f\, a \circledast_1 f^\gamma\,. \tag{2.6}$$

For simplicity, in most of the monograph, we will assume that the function a is a symmetric function of its variables, which can be interpreted as a symmetric type of interaction,

$$a(v,w) = a(w,v)\,, \qquad \forall\, v,w \in \mathbb{V}\,, \tag{2.7}$$

like e.g. in Refs. [138,170]. However, a major part of the results also holds in the asymmetric case, as in Ref. [134], see also Section 7.2.2, under some suitable additional assumptions on a. The asymmetry of the function a will be significant in Section 7.2, Chapter 11, as well as in Sections 12.3 and 13.2.3.

Under the assumption that a satisfies Eq. (2.7), we can rewrite Eq. (2.4) with Eq. (2.5) in a compact version

$$\frac{\partial}{\partial t} f(t,v) = f^{\gamma}\,(t,v)\;(a * f)\,(t,v) - f\,(t,v)\;(a * f^{\gamma})\,(t,v)\,, \tag{2.8}$$

for any $v \in \mathbb{V}$ and $t \geq 0$. By $(a * f)$ we denote the convolution–like product in $\mathbb{V}$, that is,

$$(a * f)(t,v) = \int_{\mathbb{V}} a\,(v,w)\, f\,(t,w)\; d\mu(w)\,. \tag{2.9}$$

Due to the reduction of symbols, various forms of interaction operators will be denoted by the symbol Q throughout the monograph. The exact form will follow from the context.

There exist alternative options that can result in interesting kinetic equations that have their own significance, as we will see in Chapter 9. Let

$$\begin{aligned} T_f(v,w) &= \\ &= \left(\int_{\mathbb{V}}\right)^{\gamma} A\,(w,v,w_1,\ldots,w_\gamma)\, a\,(w,w_1,\ldots,w_\gamma) \times \\ &\times\; f(t,w_1)\ldots f(t,w_\gamma)\, d\mu(w_1)\,\ldots\, d\mu(w_\gamma)\,, \qquad v,w \in \mathbb{V}\,, \end{aligned} \tag{2.10}$$

where γ here is an integer, $\gamma = 1,2,3,\ldots$ and we use the following convention

$$\left(\int_{\mathbb{V}}\right)^{k} = \underbrace{\int_{\mathbb{V}} \ldots \int_{\mathbb{V}}}_{k\,\text{times}}\,, \tag{2.11}$$

for any integer k.

Equation (2.10) leads to

$$\frac{\partial}{\partial t} f(t,v) = Q[f](t,v) \qquad t > 0\,, \quad v \in \mathbb{V}\,, \tag{2.12}$$

with

$$\begin{aligned} Q[f](t,v) &= \\ &= \left(\int_{\mathbb{V}}\right)^{(\gamma+1)} A\,(v,w,w_1,\ldots,w_\gamma)\, a\,(w,w_1,\ldots,w_\gamma) \times \\ &\times\; f(t,w) f(t,w_1)\ldots f(t,w_\gamma)\, d\mu(w)\, d\mu(w_1)\,\ldots\, d\mu(w_\gamma) \\ &-\; f(t,v) \left(\int_{\mathbb{V}}\right)^{\gamma} a\,(v,w_1,\ldots,w_\gamma) \times \\ &\times\; f(t,w_1)\ldots f(t,w_\gamma)\, d\mu(w_1)\,\ldots\, d\mu(w_\gamma)\,. \end{aligned} \tag{2.13}$$

Another possibility could be

$$\begin{aligned} T_f(v,w) &= A_0\,(w,v)\,a_0\,(v)\,f(t,v) + \\ &+ \sum_{j=1}^{\gamma} \Big(\int_{\mathbb{V}} \Big)^j A_j\,(w,v,w_1,\,\dots\,,w_j)\,a_j\,(v,w_1,\,\dots\,,w_j) \times \\ &\times\ f(t,w_1)\dots f(t,w_j)\,\mathrm{d}\mu(w_1)\,\dots\,\mathrm{d}\mu(w_j)\,, \qquad v,w\in\mathbb{V}\,. \end{aligned} \tag{2.14}$$

Equation (2.14) leads to

$$\frac{\partial}{\partial t} f(t,v) = Q[f](t,v) \qquad t\geq 0\,, \quad v\in\mathbb{V}\,, \tag{2.15}$$

with

$$\begin{aligned} Q[f](t,v) &= \int_{\mathbb{V}} A_0\,(v,w)\,a_0\,(w)\,f(t,w)\,\mathrm{d}w - a_0(v)f(t,v) + \\ &+ \sum_{j=1}^{\gamma} \Bigg(\Big(\int_{\mathbb{V}} \Big)^{(j+1)} A_j\,(v,w,w_1,\,\dots\,,w_j)\,a_j\,(w,w_1,\,\dots\,,w_j) \times \\ &\times\ f(t,w)f(t,w_1)\dots f(t,w_j)\,\mathrm{d}\mu(w)\,\mathrm{d}\mu(w_1)\,\dots\,\mathrm{d}\mu(w_j) \\ &-\ f(t,v) \Big(\int_{\mathbb{V}} \Big)^j a_j\,(v,w_1,\,\dots\,,w_j) \times \\ &\times\ f(t,w_1)\dots f(t,w_j)\,\mathrm{d}\mu(w_1)\,\dots\,\mathrm{d}\mu(w_j) \Bigg)\,. \end{aligned} \tag{2.16}$$

The terms $A_j(v,w,w_1,\dots,w_j)$ can be interpreted as the transition probabilities for the test agent to change from state w to v caused by the interaction with agents with states w_1, w_2,...,w_j, while $a_j(w,w_1,...,w_j)$ is the rate of interaction between agent with state w and agents with states w_1,...,w_j.

In Chapter 9 we will see that Equations (2.12) and (2.15), under the suitable symmetry assumption, can be directly related to the dynamics of N interacting agents at the limit $N\to\infty$. The former may be related to the interactions between γ agents, whereas the latter to the interactions between j agents, with $j=1,2,\dots,\gamma$ and $j=0$ corresponds to a stochastic change without any interaction; cf. Chapter 9. However, it should be noted that Eq. (2.15) can be reduced to Eq. (2.12) by setting a_j equal to zero for each j range from 0 to $\gamma-1$. On the other hand, due to conservative properties, Eq. (2.12) formally results in Eq. (2.15), with a suitable choice of definitions of A and a.

In the rest of the monograph, we will use the Lebesgue measure μ denoting in a standard way as

$$\mathrm{d}\,\mu(v) \;=\; \mathrm{d}v\,. \tag{2.17}$$

When referring to the equations in this section, we will always assume Eq. (2.17).

The monograph describes the basic mathematical properties of integro–differential equations which describe the distribution of the individual state of an agent of a certain population. A change in state occurs through interactions with other agents. The equations studied are non–linear and can generate very rich and diverse behaviours of solutions with significant interpretations.

The equations studied in the monograph lead to phenomena of order or disorder, depending on the situation considered. The monograph discusses whether and how it is possible to mathematically model order, that is, self–organization, or disorder, that is, disorganization, cf. Chapters 4 and 5, in life and social systems. The presented theory allows us to provide a mathematical basis for such behaviour. The blow–ups of solutions can describe some self–organization effects in a finite time, which can have a positive connotation, such as healing, cf. Section 13.2, or a negative connotation, such as society polarization. The results can be applied to processes in social sciences, economics, biology; medicine. Moreover, the importance of space–nonlocal modelling of some biological phenomena, including the invasion of cancer on the surrounding tissue, is discussed as well; see Chapter 12. Various non–local models are presented; see Chapter 12.

2.3 Opinion formation

Although the model, as mentioned above, can have many specific applications, a convenient language of interpretation is to treat an individual parameter v as an opinion of a test agent of the system, defined appropriately.

Keeping this in mind, we can follow Ref. [138] and consider the following particular societies

Assumption 1: ***A conformist society.*** *We assume*

$$T_f(w,v) = a(v,w)f^{\gamma}(t,w)\,,$$

where we assume that T_f depends on $f = f(t)$, and thus depends on t, but for simplicity of notation, we do not explicitly indicate the dependence.

The interpretation is as follows: An agent with state v is going to change its state to w with a rate that depends on the actual distribution of the latter state. The higher the probability of finding an agent with state w, the greater the tendency to change state to w. The parameter $\gamma > 0$ is related to the sensitivity of other agents to the state of a given test agent. The function $a = a(v,w)$ refers to the influence rate of agents with states v by the agent with state w. As already mentioned, the function a is termed the interaction rate or the interaction kernel.

The assumption 1 leads to Eq. (2.4).

One can consider another type of society—see Ref. [138]—in a sense, an opposite one that was called in Ref. [138]—a moderately nonconformist society.

Assumption 2: ***A moderately nonconformist society.***

$$T_f(w,v) = a(v,w)f(t,w)f^{\gamma-1}(t,v)\,,$$

In this case, the society is structured so that the test agent with the state v is likely to transition to the state w if the latter is present in the society (owing to the factor $f(t,w)$) and escapes from the former state v if it is highly represented: the factor $f^{\gamma-1}(t,v)$, with $\gamma > 1$.

Note that the following choice of $T[f(t,\,.\,)]$ leads to the same kinetic equation, again validating its name—the moderately anti conformist society.

Assumption 3. *Let*

$$T[f(t,\,.\,)](v,w) = a(v,w)f(t,w)\Big(\mathcal{M}^{\gamma-1}(t) - f^{\gamma-1}(t,v)\Big), \qquad v,w \in \mathbb{V},$$

where $\gamma > 1$ *is a given number and* $\mathcal{M}(t) = \operatorname*{ess\,sup}_{v\in\mathbb{V}} f(t,v)$.

Assumptions 2 and 3 are interesting cases as they lead to the following equation

$$\frac{\partial}{\partial t} f(t,v) = -\,Q[f](t,v), \tag{2.18}$$

where the operator Q is given by (2.5). Equation (2.18) has the form of Eq. (2.4) with the sign minus in front of the operator on the right–hand side. Thus, the forward evolution given by Eq. (2.4) is the same as the backward evolution given by Eq. (2.18).

Both cases we abbreviate as

$$\frac{\partial}{\partial t} f(t,v) = \zeta\, Q[f](t,v), \tag{2.19}$$

where $\zeta = 1$ for Eq. (2.4) and $\zeta = -1$ for Eq. (2.18), and the interaction operator Q is given by (2.5)..

Under the assumption that a is symmetric, that is, satisfies Eq. (2.7), we can rewrite Eq. (2.19) in a compact version

$$\frac{\partial}{\partial t} f(t,v) = \zeta\left(f^{\gamma}(t,v)\,(a*f)(t,v) - f(t,v)\,(a*f^{\gamma})(t,v)\right), \tag{2.20}$$

for any $v \in \mathbb{V}$ and $t \geq 0$, where $a*f$ is defined by Eq. (2.9).

The initial datum is

$$f(0,v) = \overset{\circ}{f}(v), \qquad v \in \mathbb{V}. \tag{2.21}$$

Let $\zeta = 1$. Equation (2.20) together with Eq. (2.21) leads to the following equation

$$\frac{\partial}{\partial t}\left(\frac{\exp\left(-(\gamma-1)\int_0^t (a*f^{\gamma})(s,v)\,\mathrm{d}s\right)}{f^{\gamma-1}(t,v)}\right) =$$

$$= \;-(\gamma-1)\,(a*f)(t,v)\,\exp\left(-(\gamma-1)\int_0^t (a*f^{\gamma})(s,v)\,\mathrm{d}s\right).$$

Then integrating yields

$$f(t,v) =$$

$$\left(\frac{\overset{\circ}{f}{}^{\gamma-1}(v)\,\exp\left(-(\gamma-1)\int_0^t (a*f^{\gamma})(s,v)\,\mathrm{d}s\right)}{1-(\gamma-1)\overset{\circ}{f}{}^{\gamma-1}(v)\int_0^t (a*f)(s,v)\,\exp\left(-(\gamma-1)\int_0^s (a*f^{\gamma})(r,v)\,\mathrm{d}r\right)\mathrm{d}s}\right)^{\frac{1}{\gamma-1}}. \tag{2.22}$$

2.4 Space structure

As discussed in Chapter 1, the parameter $v \in \mathbb{V}$ has a general meaning and can, in particular, contain information on the position of the agent in space. However, in some cases, as in the classical kinetic theory, the space variable can be treated separately.

Reference [137] contains an analysis of the space–dependent equation, that is, with non–local space interactions, cf. [1].

The evolution of populations at the mesoscopic scale is defined by the nonlinear integro–differential equation; see Ref. [170], taking into account the space structure

$$\frac{\partial}{\partial t} f(t,x,v) + v \cdot \frac{\partial}{\partial x} f(t,x,v) \; = \; Q[f](t,x,v) \,, \tag{2.23}$$

where

$$Q[f](t,x,v) = \int_{\mathbb{V}} \Big(T_f(t,x,w,v) f(t,x,w) - f(t,x,v)\, T_f(t,x,v,w) \Big)\, \mathrm{d}w \,,$$

where the turning rate T_f depends on $f = f(t,x,v)$ and therefore also depends on variables t and x.

Equation (2.23) can be modified to include space–non–local interactions

$$\frac{\partial}{\partial t} f(t,x,v) + v \cdot \frac{\partial}{\partial x} f(t,x,v) = Q[f](t,x,v) \,, \tag{2.24}$$

where

$$Q[f](t,x,v) = \int_{\mathbb{V}} \int_{\mathbb{D}} \Big(T_f(t,y,w,x,v) f(t,y,w) - f(t,x,v)\, T_f(t,x,v,y,w) \Big)\, \mathrm{d}y \, \mathrm{d}w \,.$$

This equation is a kind of nonlinear integro–differential equation that allows for "spatial smearing of interactions" somehow similar to the Povzner equation or an averaged Boltzmann–like equation, cf. Ref. [13, 90, 168] and references therein. Several results were obtained for the space–homogeneous version of Eq. (2.23) can be extended to the non–local equation (2.24).

Equations (2.23) and (2.24) give interesting mathematical structures that result in complex dynamics.

Assume that the function a does not depend on the space variables and is a symmetric function, that is, Eq. (2.7) is satisfied. Denoting by $(a * f)$ the convolution–like product in $\mathbb{D} \times \mathbb{V}$,

$$(a * f)(t,v) = \int_{\mathbb{V}} \int_{\mathbb{D}} a\,(v,w)\, f\,(t,y,w)\; \mathrm{d}y \, \mathrm{d}w \,, \tag{2.25}$$

we can consider the following simplified version of Eq. (2.24)

$$\begin{aligned} \tfrac{\partial}{\partial t} f(t,x,v) + v \cdot \tfrac{\partial}{\partial x} f(t,x,v) \; &= \\ = \; (a * f)\,(t,v)\, f^{\gamma}(t,x,v) - \; f(t,x,v)\; &(a * f^{\gamma})\,(t,v) \,. \end{aligned} \tag{2.26}$$

Chapter 3

Preliminary results

This section is based on Ref. [170].

Let $L^p := L^p(\mathbb{V}, \mu)$, $p \geq 1$, be the standard Banach space equipped with the norm

$$\|f\|_p = \left(\int_{\mathbb{V}} |f(v)|^p \, \mathrm{d}\mu(v) \right)^{\frac{1}{p}},$$

if $1 \leq p < \infty$, and

$$\|f\|_\infty = \operatorname*{ess\,sup}_{v \in \mathbb{V}} |f(v)|,$$

where $\operatorname{ess\,sup}_{v\in\mathbb{V}}$ is understood with respect to the measure μ. In further considerations, we consequently treat the measure μ as the Lebesgue measure and use the notation Eq. (2.17).

If $\gamma > 1$, then the reasonable setting for the solutions of Eq. (2.19) is $L^1 \cap L^\infty$ (in general) or L^∞ if additionally $\mathbb{V}$ is a bounded domain.

Throughout the monograph, χ is the standard indicator function defined as follows

$$\chi(\text{true}) = 1, \qquad \chi(\text{false}) = 0. \tag{3.1}$$

The important properties of the solution are given by the following *a priori* statements.

Proposition 1 *Let $f = f(t)$ be a solution of Eq.* (2.19). *We have*

- *For $\gamma > 1$ the solution f corresponding to a nonnegative initial datum is a priori nonnegative,*
- *The integral in $\mathbb{V}$ of the solution $f = f(t)$ is a priori conserved*

$$\int_{\mathbb{V}} f(t, v) \, \mathrm{d}v = \int_{\mathbb{V}} \overset{\circ}{f}(v) \, \mathrm{d}v,$$

where $\overset{\circ}{f}(v) = f(0, v)$ is the initial datum, $\overset{\circ}{f} \in L^1$.

Proof. The statement "*a priori*" means that the property is valid at the time of existence of the solution. By the form of Eq. (2.19), the proof follows in both cases $\zeta = 1$ and $\zeta = -1$, cf. Ref. [170]. ■

By Proposition 1 the probabilistic meaning of the solution *a priori* follows: if the initial datum $\overset{\circ}{f}$ is a probability density, then the solution is also a probability density in the time interval in which the solution exists.

In particular in the symmetric scenario, that is, where a is a symmetric function of the variables v and w, Eq. (2.19) simplifies to a trivial form for $\gamma = 1$. In fact, in this case the right–hand side of the equation is simply 0. Therefore, in this case, we can consider two nontrivial situations; cf. Ref. [170],

DOI: 10.1201/9781003537595-3

- $0 < \gamma < 1$,
- $\gamma > 1$.

However, it should be noted that the asymmetric a, that is, if Eq. (2.7) is not satisfied, for $\gamma = 1$, is no longer trivial and may also be important in the description of processes in which interactions do not exhibit symmetric properties. This case is discussed in Sections 7.2 and Chapter 11.

In the present monograph, we focus our attention primarily on the case $\gamma > 1$.

The basic assumption is the following, but in various cases it can be relaxed.

Assumption 4 *Assume that a is a symmetric function of its variables, that is, Eq. (2.7) is satisfied. Moreover*

$$a_- \le a(v,w) \le a_+ , \qquad \forall\, v,w \in \mathbb{V}, \tag{3.2}$$

where a_- and a_+ are positive constants $0 < a_- \le 1 \le a_+ < \infty$.

The following *a priori* statement, cf. Ref. [170], explains the potential possibility of a description of order, that is organization or disorder, that is, disorganization, by Eq. (2.19).

We have the following *a priori* statement; see Ref. [170].

Proposition 2 *Let*

- *a satisfy Eq. (2.7),*
- *$\vartheta : \mathbb{R}_+ \to \mathbb{R}$ is a bounded convex function,*
- *$f = f(t)$ is a solution of Eq. (2.19).*

Then f satisfies the following statement.

1. *If one of the following two conditions holds,*

   (a) $0 < \gamma < 1$ *and* $\zeta = 1$,
   (b) $\gamma > 1$ *and* $\zeta = -1$,

   then

$$\frac{\mathrm{d}}{\mathrm{d}t} \int_{\mathbb{V}} \vartheta\big(f(t,v)\big)\, \mathrm{d}v \le 0 . \tag{3.3}$$

2. *If one of the following two conditions holds,*

   (a) $\gamma > 1$ *and* $\zeta = 1$,
   (b) $0 < \gamma < 1$ *and* $\zeta = -1$,

   then

$$\frac{\mathrm{d}}{\mathrm{d}t} \int_{\mathbb{V}} \vartheta\big(f(t,v)\big) \mathrm{d}v \ge 0 . \tag{3.4}$$

Proof. The proof follows Ref. [170] and is the same in all three situations **[C]**, **[D]** and **[CD]** of Section 2.2. If $f = f(t, v)$ is a solution for $t > 0$ and $v \in \mathbb{V}$ then

$$\begin{aligned} &\frac{\mathrm{d}}{\mathrm{d}t} \int_{\mathbb{V}} \vartheta\left(f(t,v)\right) \mathrm{d}v = \\ = \;& \zeta \int_{\mathbb{V}} \int_{\mathbb{V}} a(v,w)\Big(f^{\gamma}(t,v)f(t,w) - f^{\gamma}(t,w)f(t,v)\Big)\vartheta'\left(f(t,v)\right) \mathrm{d}w\, \mathrm{d}v \\ = \;& \zeta \int_{\mathbb{V}} \int_{\mathbb{V}} a(v,w)\, f(t,v)f(t,w)\Big(f^{\gamma-1}(t,v) - f^{\gamma-1}(t,w)\Big) \times \\ \times \;& \vartheta'\left(f(t,v)\right) \mathrm{d}w\, \mathrm{d}v\,. \end{aligned}$$

Interchanging v and w variables and applying Eq. (2.7)

$$\begin{aligned} &\frac{\mathrm{d}}{\mathrm{d}t} \int_{\mathbb{V}} \vartheta\left(f(t,v)\right) \mathrm{d}v = \\ = \;& \frac{\zeta}{2} \int_{\mathbb{V}} \int_{\mathbb{V}} a(v,w)\, f(t,v)f(t,w)\Big(f^{\gamma-1}(t,v) - f^{\gamma-1}(t,w)\Big) \times \\ \times \;& \Big(\vartheta'\left(f(t,v)\right) - \vartheta'\left(f(t,w)\right)\Big) \mathrm{d}w\, \mathrm{d}v\,. \end{aligned}$$

Taking into account that the function $x \to x^{\gamma-1}$ decreases for $0 < \gamma < 1$ and increases for $\gamma > 1$, we obtain the conclusion. ■ QED

The former case, that is, when Eq. (3.3) is satisfied, refers to the entropy dissipation law, and consequently, Eq. (2.19) describes the disorganization, disorder properties, of the system—see Ref. [216]—as in classical kinetic theory of gases; see Refs. [54]. In the latter case, that is, when Eq. (3.4) holds, self–organization, that is, order properties, are observed; cf. [105, 106, 157].

The asymmetric case, that is, if Eq. (2.7) is not satisfied, is more complex, and then there is no property similar to that defined by Proposition 2. Moreover, in such a case, Eq. 2.19 with $\gamma = 1$ is no longer trivial, as we already mentioned, and may be an interesting example in modelling; cf. Sections 7.2, Chapter 11, Sections 12.3 and 13.2.3.

We state now a global in–time existence and uniqueness result for $\zeta = -1$ and $\gamma > 1$. Following Ref. [170], we formulate the result for a symmetric function a, that is, when Eq. (2.7) is satisfied, but the generalization for an asymmetric function is straightforward.

The behaviour of the solutions can refer to the following auxiliary simple ordinary differential equation that is going to be the basis for the sequence of successive approximations.

$$\frac{\mathrm{d}y}{\mathrm{d}t} = b - cy^{\gamma}\,, \qquad y\Big|_{t=0} = \mathring{y} \geq 0\,,$$

where b and c are given nonnegative continuous functions on $\mathbb{R}_+$ and $\gamma > 1$.

Then, we have, see Ref. [170],

Theorem 1 *Let*

- $\zeta = -1$, $\gamma > 1$,

- *a be measurable, symmetric (satisfying Eq. (2.7)) and nonnegative function on $\mathbb{V}^2$ such that*

$$\int_{\mathbb{V}} \sup_{w\in\mathbb{V}} a(v,w)\,dv \le a^+ ,$$

where $a^+ < \infty$ is a positive constant,

- *$\overset{\circ}{f} \in L^\infty$ be a probability density such that there exist nonnegative constants $0 \le \mathfrak{c}_m \le \mathfrak{c}_M < \infty$ satisfying*

$$\mathfrak{c}_m \le \overset{\circ}{f}(v) \le \mathfrak{c}_M \qquad \forall\, v \in \mathbb{V}. \tag{3.5}$$

Then there exists a unique solution $f \in C^1\left([0,\infty); L^1 \cap L^\infty\right)$ of Eq. (2.19) with the initial datum (2.21). Moreover, the solution $f = f(t)$, for any $t > 0$, is a density probability and satisfies

$$\mathfrak{c}_m \le f(t,v) \le \mathfrak{c}_M \qquad \forall\, t > 0 \quad \forall\, v \in \mathbb{V}. \tag{3.6}$$

Proof. The proof follows Ref. [170]. We define the sequence of successive approximations,

$$f_0(t,v) = \overset{\circ}{f}(v), \qquad \forall \quad t \ge 0, \quad v \in \mathbb{V}, \tag{3.7}$$

and

$$\frac{\partial}{\partial t} f_{n+1} = \mathcal{H}_n^+ - f_{n+1}^\gamma\, \mathcal{H}_n^- , \quad f_{n+1}\Big|_{t=0} = \overset{\circ}{f}, \qquad n = 0,1,\ldots , \tag{3.8}$$

where

$$\mathcal{H}_n^+ = f_n\,(a * f_n^\gamma)\,, \qquad \mathcal{H}_n^- = (a * f_n)\,.$$

By Eq. (3.5) we have

$$\mathfrak{c}_m \le f_0(t,v) \le \mathfrak{c}_M\,, \qquad \forall\, t > 0 \quad \forall\, v \in \mathbb{V}. \tag{3.9}$$

We want to prove that

$$\mathfrak{c}_m \le f_j(t,v) \le \mathfrak{c}_M \qquad \forall\, t > 0 \quad \forall\, v \in \mathbb{V}, \tag{3.10}$$

for all $j \in \mathbb{N}$. Assuming that Eq. (3.10) holds for $j = n$ we are going to prove it for $n+1$.
We have

$$\begin{aligned} &\tfrac{\partial}{\partial t}(f_{n+1} - \mathfrak{c}_M) = \\ &= \Big(f_n\,(a * f_n^\gamma) - \mathfrak{c}_M^\gamma\,(a * f_n)\Big) - \Big(f_{n+1}^\gamma\,(a * f_n) - \mathfrak{c}_M^\gamma\,(a * f_n)\Big). \end{aligned} \tag{3.11}$$

and

$$(a * f_n^\gamma) \le \mathfrak{c}_M^{\gamma-1}\,(a * f_n)\,. \tag{3.12}$$

Multiplying both sides of Eq. (3.11) by $\Big(f_{n+1} - \mathfrak{c}_M\Big)_+$, where

$$f_+ = \max\Big\{f\,,0\Big\},$$

and using Eq. (3.12) and Eq. (3.10) with $j = n$, we observe that both terms

$$\Big(f_{n+1} - \mathfrak{c}_M\Big)_+ \Big(f_n\,(a * f_n^\gamma) - \mathfrak{c}_M^\gamma\,(a * f_n)\Big),$$

and

$$-\left(f_{n+1}-\mathfrak{c}_M\right)_+\left(f_{n+1}^{\gamma}-\mathfrak{c}_M^{\gamma}\right)(a*f_n)$$

are non–positive, because the function $x \to x^\gamma$, $x \geq 0$, increases. Therefore, we obtain

$$\left(f_{n+1}-\mathfrak{c}_M\right)_+\frac{\partial}{\partial t}(f_{n+1}-\mathfrak{c}_M) \leq 0\,. \tag{3.13}$$

Similarly, using the inequality,

$$(a*f_n^{\gamma}) \geq \mathfrak{c}_m^{\gamma-1}(a*f_n) \tag{3.14}$$

we obtain

$$\left(\mathfrak{c}_m-f_{n+1}\right)_+\frac{\partial}{\partial t}(\mathfrak{c}_m-f_{n+1}) \leq 0\,. \tag{3.15}$$

By induction, it follows that Eq. (3.10) is satisfied for all $n \in \mathbb{N}$.

It remains to prove the convergence of the sequence $\{f_n\}_{n\in\mathbb{N}}$. To this end, we write the equation for the difference $f_{n+1}-f_n$ and multiply its both sides by the sign (the standard function sgn) of the difference. We have

$$\begin{aligned}
&\operatorname{sgn}(f_{n+1}-f_n)\,\tfrac{\partial}{\partial t}(f_{n+1}-f_n) = \\
= \;& \operatorname{sgn}(f_{n+1}-f_n)\Bigg(\tfrac{1}{2}(f_n+f_{n-1})\Big(a*(f_n^{\gamma}-f_{n-1}^{\gamma})\Big)+ \\
+ \;& \tfrac{1}{2}(f_n-f_{n-1})\Big(a*(f_n^{\gamma}+f_{n-1}^{\gamma})\Big) \\
- \;& f_n^{\gamma}\Big(a*(f_n-f_{n-1})\Big)-(f_{n+1}^{\gamma}-f_n^{\gamma})\,(a*f_n)\Bigg).
\end{aligned} \tag{3.16}$$

We see that the last term

$$-\operatorname{sgn}(f_{n+1}-f_n)\,(f_{n+1}^{\gamma}-f_n^{\gamma})\,(a*f_n)$$

is always non–positive. Furthermore, the function f_n satisfies

$$|f_n^{\gamma}(t,v)-f_{n-1}^{\gamma}(t,v)| \leq \gamma\,\mathfrak{c}_M^{\gamma-1}\,|f_n(t,v)-f_{n-1}(t,v)|\,.$$

Hence, integrating we have

$$\begin{aligned}
&\tfrac{\partial}{\partial t}\|f_{n+1}(t)-f_n(t)\|_1 \leq \\
\leq \;& \Big((\gamma+1)\,\mathfrak{c}_M^{\gamma}\int_{\mathbb{V}}\sup_{w\in\mathbb{V}} a(v,w)\,\mathrm{d}v+\mathfrak{c}_M^{\gamma}\sup_{v\in\mathbb{V}}\int_{\mathbb{V}} a(v,w)\,\mathrm{d}w\Big)\times \\
\times \;& \|f_n(t)-f_{n-1}(t)\|_1\,.
\end{aligned} \tag{3.17}$$

Therefore, for any fixed $\tau > 0$, we obtain

$$\|f_{n+1}(t)-f_n(t)\|_1 \leq \frac{\left((\gamma+2)\mathfrak{c}_M^{\gamma}a^{+}\tau\right)^n}{n!}\|f_1(t)-f_0(t)\|_1\,, \tag{3.18}$$

for all $t \in [0,\tau]$. It follows that the sequence $\{f_n(t)\}_{n\in\mathbb{N}}$ is convergent in L^1 and also in L^∞, which completes the proof. ■ QED

The following result for $\zeta = 1$ and $0 < \gamma < 1$, cf. Theorem 2.3 in Ref. [170], is similar to that of Theorem 1, except that the initial data must be separated from 0. This is because otherwise, the operator defined by the right–hand side is no longer Lipschitz continuous. The uniqueness of the solutions is not guaranteed; see Ref. [170].

Theorem 2 *Let*

- $\zeta = 1$, $0 < \gamma < 1$,
- *a be measurable, symmetric (satisfying Eq. (2.7)) and nonnegative function on $\mathbb{V}^2$ such that*
$$\int_{\mathbb{V}} \sup_{w\in\mathbb{V}} a(v,w)\,\mathrm{d}v \leq a^{+} ,$$
where $a^{+} < \infty$ is a positive constant,
- *$\overset{\circ}{f} \in L^{\infty}$ be a density probability such that there exist constants $0 < \mathfrak{c}_m \leq \mathfrak{c}_M < +\infty$*
$$\mathfrak{c}_m \leq \overset{\circ}{f}(v) \leq \mathfrak{c}_M , \qquad \forall\, v \in \mathbb{V} . \tag{3.19}$$

Then there exists a unique global solution $f \in C^1\left([0,\infty); L^1 \cap L^{\infty}\right)$ of Eq. (2.19) with the initial datum Eq. (2.21). Moreover, the solution $f = f(t)$, for any $t > 0$, is a density probability and satisfies
$$\mathfrak{c}_m \leq f(t,v) \leq \mathfrak{c}_M , \qquad \forall\, v \in \mathbb{V} . \tag{3.20}$$

Proof. The proof follows Ref. [170]. Similarly, as in the proof of Theorem 1, we define the sequence of successive approximations,
$$f_0(t,v) = \overset{\circ}{f}(v) , \qquad \forall \quad t \geq 0, v \in \mathbb{V} , \tag{3.21}$$
and
$$\frac{\partial}{\partial t} f_{n+1} = \mathfrak{H}_n^{+} - f_{n+1}\,\mathfrak{H}_n^{-} , \quad f_{n+1}\Big|_{t=0} = \overset{\circ}{f} , \qquad n = 0, 1, \ldots , \tag{3.22}$$
where
$$\mathfrak{H}_n^{+} = f_n^{\gamma}\,(a * f_n) , \qquad \mathfrak{H}_n^{-} = (a * f_n^{\gamma}) .$$
As in the proof of Theorem 1, by induction we prove that
$$\mathfrak{c}_m \leq f_n(t,v) \leq \mathfrak{c}_M \qquad \forall\, t > 0 \quad \forall\, v \in \mathbb{V} , \tag{3.23}$$
for all $n \in \mathbb{N}$, using the inequalities
$$\mathfrak{c}_m^{1-\gamma}\,(a * f_n^{\gamma}) \leq (a * f_n) \leq \mathfrak{c}_M^{1-\gamma}\,(a * f_n^{\gamma}) . \tag{3.24}$$
Then as in Eq. (3.17) we get
$$\begin{aligned} \tfrac{\partial}{\partial t}\, \|f_{n+1}(t) - f_n(t)\|_1 \leq & \Bigg(\mathfrak{c}_M^{\gamma} \textstyle\int_{\mathbb{V}} \sup_{w\in\mathbb{V}} a(v,w)\,\mathrm{d}v + \\ + & \tfrac{\mathfrak{c}_M}{\mathfrak{c}_m^{1-\gamma}}\,\gamma \sup_{v\in\mathbb{V}} \textstyle\int_{\mathbb{V}} a(v,w)\,\mathrm{d}w + \tfrac{\mathfrak{c}_M}{\mathfrak{c}_m^{1-\gamma}}\,\gamma \int_{\mathbb{V}} \sup_{w\in\mathbb{V}} a(v,w)\,\mathrm{d}v \Bigg) \times \\ \times & \|f_n(t) - f_{n-1}(t)\|_1 . \end{aligned} \tag{3.25}$$

Therefore, for any fixed $\tau > 0$, we have

$$\|f_{n+1}(t) - f_n(t)\|_1 \leq \frac{\left(\left(c_M^\gamma + \frac{2\,c_M}{c_m^{1-\gamma}}\right)a^+\tau\right)^n}{n!}\,\|f_1(t) - f_0(t)\|_1\,, \tag{3.26}$$

for all $t \in [0, \tau]$. It follows that the sequence $\{f_n(t)\}_{n\in\mathbb{N}}$ is convergent in L^1 and also in L^∞, which completes the proof. ■ QED

Now, cf. Ref. [170], we state the global existence result without uniqueness for $\zeta = -1$ and $0 < \gamma < 1$. The behaviour of the solutions can refer to the following auxiliary simple ordinary differential equation that is going to be a base for successive approximations.

$$\frac{\mathrm{d}\,y}{\mathrm{d}\,t} = by - cy^\gamma\,, \qquad y\Big|_{t=0} = \overset{\circ}{y} \geq 0\,,$$

where b and c are nonnegative continuous functions on $\mathbb{R}_+$. For $0 < \gamma < 1$, the solution may not be unique. However, we can extend the solution by 0, for example, the solution for constant b and c can have the form

$$y(t) = \begin{cases} \left(\frac{c}{b} + \left(\overset{\circ}{y}^{\,1-\gamma} - \frac{c}{b}\right)\exp\left((1-\gamma)\,b\,t\right)\right)^{\frac{1}{1-\gamma}} & \text{for} \quad t < \tau\,, \\ 0 & \text{for} \quad t \geq \tau\,, \end{cases} \tag{3.27}$$

where

$$\tau = \begin{cases} -\frac{\log\left(1 - \frac{b}{c}\,\overset{\circ}{y}^{\,1-\gamma}\right)}{(1-\gamma)\,b} & \text{if} \quad \overset{\circ}{y} < \left(\frac{c}{b}\right)^{\frac{1}{1-\gamma}}\,, \\ +\infty & \text{if} \quad \overset{\circ}{y} \geq \left(\frac{c}{b}\right)^{\frac{1}{1-\gamma}} \end{cases}\,.$$

We have the following global existence result; see Ref. [170].

Theorem 3 *Let*

- $\mathbb{V}$ *be a bounded domain,*
- $\zeta = -1$, $0 < \gamma < 1$,
- a *satisfy Assumption 4,*
- *the initial datum* $\overset{\circ}{f}$ *given by Eq.* (2.21) *satisfies* $\overset{\circ}{f} \in L^\infty$.

Then, for any $t > 0$, *there exists a solution* $f = f(t)$ *of Eq.* (2.19) *in* L^∞.

***Proof*.** The proof follows the proof of Theorem 2.4 in Ref. [170], where moreover, the estimations of time and domain for which the solution vanishes are given.

We define the following sequence of successive approximations,

$$f_0(t, v) = \overset{\circ}{f}\,(v)\,, \qquad \forall \quad t \geq 0, \quad v \in \mathbb{V}\,, \tag{3.28}$$

and f_{n+1}, $n = 0, 1, 2, \ldots$ is a nonnegative function that satisfies

$$\begin{aligned} &\tfrac{\partial}{\partial\,t} f_{n+1} = f_{n+1}\,(a * f_n^\gamma) - f_{n+1}^\gamma\,(a * f_n)\,, \\ &f_{n+1}\Big|_{t=0} = \overset{\circ}{f}\,, \qquad n = 0, 1, \ldots\,. \end{aligned} \tag{3.29}$$

The function $\overset{\circ}{f} \in L^\infty$ is nonnegative. Assuming that $f_n \in C^0\,(0,\infty\,;L^\infty)$ is nonnegative we have

$$\frac{\partial}{\partial t} f_{n+1}^{1-\gamma} = (1-\gamma)\left(f_{n+1}^{1-\gamma}\,(a * f_n^\gamma) - (a * f_n)\right).$$

The nonnegative solution satisfies

$$\frac{\partial}{\partial t} f_{n+1}^{1-\gamma} \le (1-\gamma)\left(a_+ \|f_n\|^\gamma_{L^\infty(0,\infty\,;L^\infty)}\, f_{n+1}^{1-\gamma} - a_-\right).$$

By the Grönwall inequality we obtain

$$f_{n+1}^{1-\gamma} \in C^0\left([0,\infty)\,;L^\infty\right).$$

Therefore f_{n+1} is nonnegative and

$$f_{n+1} \in C^0\left([0,\infty)\,;\,L^\infty\right).$$

The convergence of the sequence $\{f_n\}_{n\in\mathbb{N}}$ in some interval of time $[0,\tau_1]$, $\tau_1 > 0$, follows by the standard contraction argument.

The function $\vartheta(x) = x^\gamma$, $x \ge 0$, $0 < \gamma < 1$, is concave. Thus, using the argument of Proposition 2, we show that if $f = f(t,x)$ is a solution, then

$$\int_{\mathbb{V}} f^\gamma(t,v)\,dv \le \int_{\mathbb{V}} \overset{\circ}{f}{}^\gamma\,(v)\,dv := \overset{\circ}{\mathfrak{c}}\,, \qquad t \ge 0\,,$$

Therefore, we have

$$\frac{\partial}{\partial t} f(t,v) \le a_+\ \overset{\circ}{\mathfrak{c}}\ f(t,v)\,,$$

and the statement of Theorem 3 follows. ■ QED

The most interesting case is $\zeta = 1$ and $\gamma > 1$. In the general situation, only a local (in time) result is possible. This case will be discussed in detail in the following chapter.

We have the following local result; see Ref. [170].

Theorem 4 *Let*

- $\zeta = 1$, $\gamma > 1$,
- *a satisfy Assumption 4,*
- *$\overset{\circ}{f} \in L^\infty$ be a density probability.*

Then there exist $\tau > 0$ and a unique solution of Eq. (2.19) *$f = f(t)$ in $C^1\left([0,\tau)\,;L^\infty\right)$. Moreover, the solution $f = f(t)$ is a probability density for all $t \in [0,\tau)$.*

Proof. The operator given on the right-hand side of Eq. (2.19) is Lipschitz–continuous. Applying Proposition 1, we obtain the conclusion, cf. Ref. [170], Theorem 2.5. ■ QED

As we will see in Section 4 the general behaviour of the solution of Eq. (2.23) for $\gamma > 1$ can be quite complex. This provides interesting possibilities for describing phenomena in the life and social sciences and, on the other hand, leads to interesting mathematical considerations.

In contrast to Eq. (2.23) with $\gamma > 1$, Eq. (2.23) with $\gamma = 1$ when an asymmetric kernel is assumed, as well as Eqs. (2.12) and (2.15) result in the global existence–uniqueness of solutions.

Proposition 3 *Let*

- $\gamma = 1$,
- $a \in L^\infty(\mathbb{V} \times \mathbb{V})$,
- $\overset{\circ}{f}$ *be a probability density.*

Then, for any $t > 0$*, a solution* $f = f(t)$ *to Eq. (2.4), corresponding to the initial datum* $\overset{\circ}{f}$*, exists and is unique in* L^1*. The solution preserves positivity and the* L^1*–norm; that is, it is a probability density, for any* $t > 0$*.*

Proof. The proof follows in a standard way by the Lipschitz continuity of the operator on the right–hand side of Eq. (2.4) together with a priori nonnegativity and the preservation of L^1–norm of the solution—cf. Proposition 1. ■ QED

Assumption 5 *Let* γ *be an integer and*

$$\begin{aligned} & A_j \geq 0\,, \qquad a_j \geq 0\,, \qquad a_j \in L^\infty\left(\mathbb{V}^{j+1}\right)\,, \\ & \int_{\mathbb{V}} A_j\left(v, w, w_1, \ldots, w_j\right) dv = 1 \quad \text{for all} \quad (w, w_1, \ldots, w_j) \in \mathbb{V}^{j+1} \\ & \text{such that} \quad a_j\left(w, w_1, \ldots, w_j\right) > 0\,, \qquad \forall\, j = 1, \ldots, \gamma\,. \end{aligned} \tag{3.30}$$

Under Assumption 5 we obtain the global existence and uniqueness result for Eq. (2.15).

Proposition 4 *Let*

- *Assumption 6.2 be satisfied,*
- $\overset{\circ}{f}$ *be a probability density.*

Then, for any $t > 0$*, the solution* $f = f(t)$ *to Eq. (2.15), corresponding to the initial datum* $\overset{\circ}{f}$*, exists and is unique in* L^1*. The solution preserves both positivity and the* L_1*–norm; that is, it is a probability density for* $t > 0$*.*

Corollary 1 *The solutions in Propositions 3 and 4 are in* L^∞ *on every compact interval* $[0, \tau]$*,* $\tau > 0$*, provided that* $\overset{\circ}{f} \in L_\infty$*.*

Proof. The proofs of Proposition 4 and Corollary 1 are standard and are based on the Lipschitz property in L^1 and L^∞, respectively. ■

Chapter 4

Constant interaction rate

In this chapter, we consider Case [**C**] as introduced in Section 2.2. Hence, μ is the Lebesgue measure, and we write $\mathrm{d}v$ instead of $\mathrm{d}\mu(v)$. Additionally, we assume $\gamma > 1$ and $\zeta = 1$ and therefore deal with Eq. (2.4). We follow Ref. [136].

The case where a does not depend on its variables, that is, a is a (positive) constant, represents the simplest situation. Although the potential applications may be somewhat limited in this case, it is possible to construct a comprehensive mathematical framework for understanding solution behaviour; see Ref. [136]. This simple case can be a starting point for studying more realistic and complex cases where the theory is more demanding; cf. Ref. [137, 139, 140].

Assuming then that the kernel is constant and without loss of generality simply $a \equiv 1$, we obtain the following particular version of Eq. (2.4) and Eq. (2.19) with $\zeta = 1$

$$\frac{\partial}{\partial t} f(t,v) = f^{\gamma}(t,v)\,\|f(t,.)\|_1 - f(t,v)\,\|f(t,.)\|_{\gamma}^{\gamma}, \qquad t \geq 0,\ v \in \mathbb{V},$$

and then, referring to Proposition 1, we have

$$\frac{\partial}{\partial t} f(t,v) = f^{\gamma}(t,v) - f(t,v)\,\|f(t,.)\|_{\gamma}^{\gamma}, \qquad t \geq 0,\ v \in \mathbb{V}, \tag{4.1}$$

The initial datum is defined by Eq. (2.21).

Equation (4.1) has an interesting mathematical structure. In fact, the right–hand side of the equation is in the form of a difference of two terms. The former ("positive" or "gain") can produce blow–ups of the solutions in a finite time, whereas the second ("negative" or "loss") can prevent them. Thus, the operator on the right–hand side expresses the competition between these two terms. This type of situation often occurs in mathematics and is an interesting basis for mathematical ideas.

With the natural interpretation that blow–ups of solutions in a finite time can determine society's self–organization, we obtain an interesting possibility in modelling social behaviour. An obvious question is what factors could lead to this kind of self–organization. When can we obtain self–organization and when, on the contrary, the system remains disordered in some sense? As we shall see, as expected, the behaviour of the solutions is heavily influenced by the initial data. However, what is particularly noteworthy is that the corresponding mathematical conditions possess a compelling interpretation, even in the simplified scenario of a constant kernel.

Following Ref. [136] and keeping in mind Eq. (2.22) we denote

$$F(t) = \int_0^t \|f(s,.)\|_{\gamma}^{\gamma} \mathrm{d}s$$

and observe that

$$\frac{\mathrm{d}}{\mathrm{d}t} F(t) = \frac{1}{\exp\Big(\gamma F(t)\Big)} \int_{\mathbb{V}} \frac{\overset{\circ}{f}{}^{\gamma}(v)}{\Big(1 - (\gamma - 1)\,\overset{\circ}{f}{}^{\gamma-1}(v)\,\varphi(t)\Big)^{\frac{\gamma}{\gamma-1}}} \mathrm{d}v\,, \tag{4.2}$$

DOI: 10.1201/9781003537595-4

where

$$\varphi(t) = \int_0^t \frac{1}{\exp\Big((\gamma - 1)\, F(s)\Big)}\, \mathrm{d}s\,. \tag{4.3}$$

The objective is to substitute the equation into a suitable ordinary differential equation (ODE) that contains the information of possible blow-ups and then apply the standard theory of ODEs.

By

$$\frac{\mathrm{d}}{\mathrm{d}t}\varphi(t) = \frac{1}{\exp\Big((\gamma-1)\, F(t)\Big)}\,,$$

we obtain

$$\frac{\mathrm{d}^2}{\mathrm{d}^2 t}\varphi(t) = -(\gamma - 1)\frac{1}{\exp\Big((\gamma-1)\, F(t)\Big)}\frac{\mathrm{d}}{\mathrm{d}t}F(t)\,.$$

Upon integration, we derive the following quite complex ODE for φ

$$\frac{\mathrm{d}}{\mathrm{d}t}\varphi(t) = \Phi\big(\varphi(t)\big)\,, \qquad \varphi(0) = 0\,, \tag{4.4}$$

that involves the initial datum $\overset{\circ}{f}$, where

$$\Phi(\varphi) = \frac{1}{\left(\displaystyle\int_{\mathbb{V}} \frac{\overset{\circ}{f}(v)}{\left(1-(\gamma-1)\,\overset{\circ}{f}^{\,\gamma-1}(v)\,\varphi\right)^{\frac{1}{\gamma-1}}}\,\mathrm{d}v\right)^{\gamma-1}}\,. \tag{4.5}$$

We observe that the function $\varphi = \varphi(t)$ is both increasing and concave. The condition for the blow-up at $\tau > 0$ is given by the following equality

$$(\gamma - 1)\,\|\, \overset{\circ}{f}\,\|_\infty^{\gamma-1}\,\varphi(\tau) = 1\,. \tag{4.6}$$

Equation (4.4) fully determines the conditions of blow-ups versus the global existence of solutions to Eq. (4.1). The following set plays a crucial role in this context.

$$\mathbb{W} = \Big\{ v \in \mathbb{V}: \quad \overset{\circ}{f}(v) = \|\, \overset{\circ}{f}\,\|_\infty \Big\}\,.$$

If the set $\mathbb{W}$ is large in the sense that $|\mathbb{W}| > 0$, where $|\,.\,|$ is the Lebesgue measure, then blow-ups in a finite time do not occur—an existence-uniqueness result of the solution, corresponding to the initial datum $\overset{\circ}{f}$, globally in time is guaranteed; see Theorem 5 below.

Condition $|\mathbb{W}| > 0$ can be understood in the following manner: when there is a significant number of dominant states, alignment does not occur in a finite time.

We have the following global existence–uniqueness result; see Ref. [136], Theorem 1.

Theorem 5 *Let*

- $\overset{\circ}{f} \in L^\infty$ *is a probability density on* $\mathbb{V}$,
- $|\mathbb{W}| > 0\,.$

Then, for any $t > 0$, there exists a unique solution $f = f(t)$ in L^∞ of Eq. (4.1) corresponding to (2.21). The solution $f = f(t)$, for any $t > 0$, is a probability density.

***Proof*.** The proof follows; see Ref. [136], by the comparison equation method. The aim is to achieve an inequality of type $\frac{d}{dt}\varphi(t)\ (\ldots) \leq 1$. We have

$$\frac{d}{dt}\varphi(t)\left(|\mathbb{W}|\,\|\overset{\circ}{f}\|_\infty\left(1-(\gamma-1)\,\|\overset{\circ}{f}\|_\infty^{\gamma-1}\varphi(t)\right)^{\frac{1}{1-\gamma}}+\right.$$
$$\left.+\int\limits_{W'}\overset{\circ}{f}(v)\left(1-(\gamma-1)\,\overset{\circ}{f}{}^{\gamma-1}(v)\,\varphi(t)\right)^{\frac{1}{1-\gamma}}dv\right)^{\gamma-1}=1$$

where $\mathbb{W}' = \mathbb{V}\setminus\mathbb{W}$. The derivative $\frac{d}{dt}\varphi(t)$ is nonnegative and we have

$$(\gamma-1)\left\|\overset{\circ}{f}\right\|_\infty^{\gamma-1}\varphi(t)\leq 1\,,$$

cf. Eq. (4.6). Thus, the desired inequality is established

$$\frac{d}{dt}\varphi(t)\left(|\mathbb{W}|\,\|\overset{\circ}{f}\|_\infty\left(1-(\gamma-1)\,\|\overset{\circ}{f}\|_\infty^{\gamma-1}\varphi(t)\right)^{\frac{1}{1-\gamma}}\right)^{\gamma-1}\leq 1\,.$$

We consider the Cauchy problem for the following comparison equation

$$\frac{d}{dt}\Psi=\left(|\mathbb{W}|\,\|\overset{\circ}{f}\|_\infty\right)^{1-\gamma}\left(1-(\gamma-1)\,\|\overset{\circ}{f}\|_\infty^{\gamma-1}\Psi\right),\qquad \Psi(0)=0\,,$$

and obtain

$$\varphi(t)\leq\Psi(t),\qquad t\geq 0\,,$$

where Ψ is given by

$$\Psi(t)=(\gamma-1)^{-1}\|\overset{\circ}{f}\|_\infty^{1-\gamma}\left(1-e^{-t\,(\gamma-1)|\mathbb{W}|^{1-\gamma}}\right). \tag{4.7}$$

Hence, the global existence follows. ■ QED

The case when the set $\mathbb{W}$ is small, in the sense that $|\mathbb{W}| = 0$, is more complex and can lead to various possible scenarios. Function Φ, given by Eq. (4.5), is defined on $[0, \varphi_+)$, where

$$\varphi_+=\frac{\|\overset{\circ}{f}\|_\infty^{1-\gamma}}{\gamma-1}\,.$$

The function Φ is continuous and monotone. Let

$$\Phi(\varphi_+):=\lim_{\varphi\uparrow\varphi_+}\Phi(\varphi)\,.$$

We have, see Theorem 2 in Ref. [136],

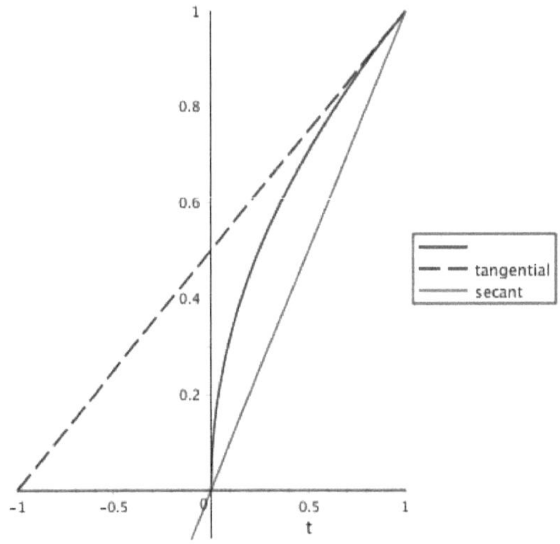

FIGURE 4.1: Illustration of the function $\varphi = \varphi(t)$, secant and tangential at (τ, φ_+).

Theorem 6 *Let*

- $\overset{\circ}{f} \in L^\infty$ *be a probability density,*
- $|\mathbb{W}| = 0$.

Then

1. *If* $\Phi(\varphi_+) > 0$, *then a blow–up in a finite time of the solution Eq.* (4.1) *takes place,*
2. *If* $\Phi(\varphi_+) = 0$, *then*
   (a) *If* $\int_0^{\varphi_+} \frac{1}{\Phi(\varphi)}\,\mathrm{d}\varphi < \infty$ *then a blow–up in a finite time of the solution Eq.* (4.1) *take place,*
   (b) *If* $\int_0^{\varphi_+} \frac{1}{\Phi(\varphi)}\,\mathrm{d}\varphi = \infty$ *then for each* $\tau > 0$ *there exists a solution of the solution Eq.* (4.1) *on* $[0, \tau]$.

Proof. Cf. Ref. [136]. We consider first $\Phi(\varphi_+) > 0$. We observe that $\Phi'(\varphi) < 0$ and, therefore, the solution to Eq. (4.4) is increasing and concave.

The tangent of the straight line passing through the points $(0, 0)$ and (τ, φ_+) is bigger than the tangent of the tangential to the curve defined by $\varphi = \varphi(t)$ at the point (τ, φ_+)—see Figure 4.1. Therefore, the blow–up time τ satisfies

$$\frac{\varphi_+}{\Phi(0)} \leq \tau \leq \frac{\varphi_+}{\Phi(\varphi_+)}\,.$$

Now $\Phi(\varphi_+) = 0$ and then φ_+ is an equilibrium. One can apply the standard theory of ODE to obtain the statement, cf. Ref. [136]. ■ QED

Let $\overset{\circ}{f}_{\bullet}(v) = \frac{\overset{\circ}{f}(v)}{\|\overset{\circ}{f}\|_\infty}$. It is easy to see that Conditions

$$\Phi(\varphi_+) > 0 \qquad \text{and} \qquad \Phi(\varphi_+) = 0$$

are equivalent to

$$\overset{\circ}{f}_{\bullet}(1-\overset{\circ}{f}_{\bullet}{}^{\gamma-1})^{\frac{1}{1-\gamma}} \in L^1 \qquad \text{and} \qquad \overset{\circ}{f}_{\bullet}(1-\overset{\circ}{f}_{\bullet}{}^{\gamma-1})^{\frac{1}{1-\gamma}} \notin L^1 ,$$

respectively.

Item 2 of Theorem 6 can now be formulated in the form as follows; see Corollary 2.1 in Ref. [136].

Corollary 2 *Let*

- $\overset{\circ}{f} \in L^\infty$ *be a probability density,*
- $|\mathbb{W}| = 0$,
- $$\frac{\overset{\circ}{f}_{\bullet}}{\left(1-\overset{\circ}{f}_{\bullet}{}^{\gamma-1}\right)^{\frac{1}{\gamma-1}}} \notin L^1 .$$

Then

1. *If*
$$\int_0^1 \left(\int_{\mathbb{V}} \frac{\overset{\circ}{f}_{\bullet}(v)}{\left(1-\left(\overset{\circ}{f}_{\bullet}\right)^{\gamma-1}(v)\, y\right)^{\frac{1}{\gamma-1}}} \mathrm{d}v \right)^{\gamma-1} \mathrm{d}y < \infty ,$$
then a blow-up in a finite time of the solution Eq. (4.1) *takes place.*
2. *If*
$$\int_0^1 \left(\int_{\mathbb{V}} \frac{\overset{\circ}{f}_{\bullet}(v)}{\left(1-\left(\overset{\circ}{f}_{\bullet}\right)^{\gamma-1}(v)\, y\right)^{\frac{1}{\gamma-1}}} \mathrm{d}v \right)^{\gamma-1} \mathrm{d}y = \infty ,$$
then, for any time $\tau > 0$, *there exists a solution of Eq.* (4.1) *on* $[0, \tau]$.

Let, in addition,

$$g(v) := \left(\overset{\circ}{f}_{\bullet}\right)^{2-\gamma}(v) \left(1-\left(\overset{\circ}{f}_{\bullet}\right)^{\gamma-1}(v)\right)^{\frac{\gamma-2}{\gamma-1}}$$

and

$$h(v) := -\log\left(1-\left(\overset{\circ}{f}_{\bullet}\right)^{\gamma-1}(v)\right) .$$

1. *If* $1 < \gamma < 2$ *and* $g \in L^1$ *then a blow-up of the solution to Eq.* (4.1) *occurs in finite time;*

2. *If* $1 < \gamma \leq 2$ *and* $h \notin L^1$ *then the solution to Eq.* (4.1) *is global in time,*

3. *If* $\gamma \geq 2$ *and* $h \in L^1$ *then a blow–up of the solution Eq.* (4.1) *takes place in finite time.*

Proof. It follows

$$\int_0^{\varphi_+} (\Phi(\varphi))^{-1}\, \mathrm{d}\varphi =$$

$$= \int_0^{(\gamma-1)^{-1}\|\overset{\circ}{f}\|_\infty^{1-\gamma}} \left(\int_{\mathbb{V}} \overset{\circ}{f}(v)\Big(1-(\gamma-1)\,\overset{\circ}{f}^{\,\gamma-1}(v)\,\varphi\Big)^{\frac{1}{1-\gamma}} \mathrm{d}v \right)^{\gamma-1} \mathrm{d}\varphi\,.$$

Changing the variable

$$y := (\gamma-1)\left\|\overset{\circ}{f}\right\|_\infty^{\gamma-1} \varphi\,,$$

we obtain

$$\int_0^{\varphi_+} (\Phi(\varphi))^{-1}\, \mathrm{d}\varphi =$$

$$= (\gamma-1)^{-1} \int_0^1 \left(\int_{\mathbb{V}} \overset{\circ}{f}_{\bullet}(v) \left(1 - \left(\overset{\circ}{f}_{\bullet}\right)^{\gamma-1}(v)\, y \right)^{\frac{1}{1-\gamma}} \mathrm{d}v \right)^{\gamma-1} \mathrm{d}y\,.$$

Thus, by Theorem 6, the first part of the statement follows. To prove the second part, we use the first part and Hölder's inequality. ■ QED

We can formulate now the following example, cf. Ref. [136],

Example 1 *Let* $\alpha > 0$ *and* $\overset{\circ}{f}_{\bullet}(v) = 1 - \exp\left(-\frac{1}{|v|^\alpha}\right)$ *on* $\mathbb{V} = (-1,1)\setminus\{0\}$. *We have*

$$h \in L^1 \qquad \Leftrightarrow \qquad \alpha < 1\,.$$

Therefore

- *If* $\alpha = \frac{1}{2}$ *then a blow–up occurs in finite time,*
- *If* $\alpha = 2$, *then the solution is global.*

According to Theorems 3.1 and 3.2 in Ref. [137], the following large–time asymptotic result holds.

Theorem 7 *Let* $\overset{\circ}{f} \in L^\infty$ *be a probability density on* $\mathbb{V}$.

1. *If* $|\mathbb{W}| > 0$, *then the unique global solution* f *of Eq.* (4.1), *corresponding to the initial datum* $\overset{\circ}{f}$, *satisfies*

$$\lim_{t\to\infty} f(t,v) = \frac{1}{|\mathbb{W}|}\,\chi(v\in\mathbb{W})\,,$$

where χ *is defined by Eq.* (3.1).

2. *If $|\mathbb{W}| = 0$ and f are the unique solution of Eq. (4.1) in $C^1(0, \tau_+; L^\infty)$, corresponding to the initial datum $\overset{\circ}{f}$, where τ_+ is either $\tau_+ = \infty$ or $\tau_+ = \tau < \infty$ (τ is a blow–up time).*

   (a) *If $\Phi(\varphi_+) = 0$ then*

$$\lim_{t\to\tau_+} f(t,v) = 0 \qquad \text{for} \quad v \notin \mathbb{W},$$

   (b) *If $\Phi(\varphi_+) > 0$ then*

$$\lim_{t\to\tau_+} f(t,v) = \frac{\overset{\circ}{f}_\bullet(v)}{\left(1-\left(\overset{\circ}{f}_\bullet\right)^{\gamma-1}(v)\right)^{\frac{1}{\gamma-1}} \int_{\mathbb{V}} \frac{\overset{\circ}{f}_\bullet(w)}{\left(1-\left(\overset{\circ}{f}_\bullet\right)^{\gamma-1}(w)\right)^{\frac{1}{\gamma-1}}}\, dw},$$

where $\overset{\circ}{f}_\bullet(v) = \dfrac{\overset{\circ}{f}(v)}{\|\overset{\circ}{f}\|_\infty}$.

Proof. The proof follows Ref. [137], Theorems 3.1 and 3.2. By Eq. (2.22) we have

$$f(t,v) = \frac{\overset{\circ}{f}(v)}{\exp\left(F(t)\right)\left(1-(\gamma-1)\,\overset{\circ}{f}^{\gamma-1}(v)\,\varphi(t)\right)^{\frac{1}{\gamma-1}}}. \tag{4.8}$$

Using Eqs. (4.3), (4.4), and (4.5) yields

$$F(t) = \ln\left(\int_{\mathbb{V}} \frac{\overset{\circ}{f}(v)}{\left(1-(\gamma-1)\,\overset{\circ}{f}^{\gamma-1}(v)\varphi(t)\right)^{\frac{1}{\gamma-1}}}\, dv\right).$$

Substitute $\phi(t)$ into Eq. (4.8) we obtain

$$f(t,v) =$$

$$= \frac{\overset{\circ}{f}(v)}{\left(1-(\gamma-1)\overset{\circ}{f}^{\gamma-1}(v)\varphi(t)\right)^{\frac{1}{\gamma-1}} \int_{\mathbb{V}} \frac{\overset{\circ}{f}(w)}{\left(1-(\gamma-1)\overset{\circ}{f}^{\gamma-1}(w)\,\varphi(t)\right)^{\frac{1}{\gamma-1}}}\, dw},$$

that express f as a function of φ.

Now we decompose $\mathbb{V} = \mathbb{W} \cup \mathbb{W}'$ and note that $\varphi(t) \to \varphi_+$ for $t \to \infty$. This concludes Item 1. Item 2 follows analogously. ■ QED

Therefore, we can conclude that if $|\mathbb{W}| = 0$ and $\Phi(\varphi_+) = 0$ then the solution $f(t,v)$ tends to a kind of the Dirac delta distribution concentrated on $\mathbb{W}$, as $t \to \tau_+$.

The case considered in this chapter is simple and has a rather limited role from the point of view of describing phenomena in the life and social sciences. On the other hand, in this case, it is possible to thoroughly see what the behaviour of the solutions depends on. In this situation, a comprehensive understanding of the solutions' behaviour can be achieved.

When the number of dominant states of the initial datum is large, that is, if the set $\mathbb{W}$ has a positive Lebesgue measure, $|\mathbb{W}| > 0$, then no blow–up can occur in a finite time. In fact, the solution is global in time; see Theorem 5. We can interpret this result by saying that no self–organization, in a finite time, is possible. We understand self–organization as a kind of alignment of states, cf. Section 13.2. On the other hand, if the set $\mathbb{W}$ is small in the sense of the Lebesgue measure, that is, $|\mathbb{W}| = 0$, we can specify conditions for the occurrence of a blow–up in finite time; see Theorem 6. These conditions can be expressed in easy–to–verify formulas; see Corollary 2 and Example 1. The behaviour of solutions in the case of a blow–up determines Theorem 7, item 2a.

The methods presented above can be easily adapted; see Remark 3.2 in Ref. [137], to a space–involving equation that is a version of Eq. (2.26) corresponding to the constant interaction kernel $a = 1$, as before for simplicity.

$$\begin{aligned} &\tfrac{\partial}{\partial t} f(t,x,v) + v \cdot \tfrac{\partial}{\partial x} f(t,x,v) = \\ &= \quad f^{\gamma}(t,x,v) - f(t,x,v) \int\limits_{\mathbb{V}}\int\limits_{\mathbb{D}} f^{\gamma}(t,y,w)\, \mathrm{d}y\, \mathrm{d}w\,, \end{aligned} \tag{4.9}$$

where $x \in \mathbb{D}$ is interpreted as the space variable, $\mathbb{D}$ is a domain in $\mathbb{R}^{\mathfrak{n}}$. Here, $\mathbb{D}$ can be $\mathbb{R}^{\mathfrak{n}}$ or $\mathbb{T}^{\mathfrak{n}}$ – the $\mathfrak{n}$–dimensional torus that corresponds to periodic boundary conditions or a general domain with the boundary conditions as in the paper by S. Kaniel and M. Shinbrot; Ref. [116]. Still referring to this paper Ref. [116] we can rewrite Eq. (4.9) along characteristics

$$\tfrac{\partial}{\partial t} f^{\sharp}(t,x,v) = \left(f^{\gamma}\right)^{\sharp}(t,x,v) - f^{\sharp}(t,x,v) \int\limits_{\mathbb{V}}\int\limits_{\mathbb{D}} f^{\gamma}(t,y,w)\, \mathrm{d}y\, \mathrm{d}w\,, \tag{4.10}$$

Function $f^{\sharp}$ is considered along characteristics that is, for example, if $\mathbb{D} = \mathbb{R}^{\mathfrak{n}}$ or $\mathbb{D} = \mathbb{T}^{\mathfrak{n}}$ we have

$$f^{\sharp}(t,x,v) = f(t, x + tv, v)\,.$$

The solution of Eq. (4.10) is called the mild solution of Eq. (4.9).
If

$$0 \le \mathring{f} \in L^{\infty}(\mathbb{D} \times \mathbb{V})$$

is a probability density and $|\widetilde{\mathbb{W}}| > 0$, where $\widetilde{\mathbb{W}} \subset \mathbb{D} \times \mathbb{V}$ is defined analogously to $\mathbb{W}$, then, for any $t > 0$, there exists a unique mild solution $f = f(t)$ of Eq. (4.9) in

$$L^{1}(\mathbb{D} \times \mathbb{V}) \cap L^{\infty}(\mathbb{D} \times \mathbb{V})$$

and the solution is nonnegative. Moreover, the case $|\widetilde{\mathbb{W}}| = 0$ leads to conditions analogous to those before.

Further remarks on spatially inhomogeneous equations can be found in Section 5.2.

Chapter 5

Non–constant kernel of interaction

Reference [137] contains generalizations of the results of Chapter 4 to the case where the rate a is not necessarily constant. The conditions that guarantee the existence of solutions globally in time versus their potential blow–ups in finite time are formulated in terms of complicated dependencies between the rate a and the initial data $\overset{\circ}{f}$; see the definition of the set $\mathbb{W}$ below—see Eq. (5.3). It would be difficult to give a satisfactory interpretation in applications of the formulated conditions. Therefore, the above results should be treated as preliminary steps for further research.

The rate a that is not constant is, of course, more complex to study than the constant rate, and we are still far from understanding the possible behaviour of solutions in the fully general case. Reference [137] also contains the analysis of space–dependent equations with non–local space interactions.

However, several results, obtained for the space–homogeneous version of Eq. (2.23), can be extended to the non–local equation (2.24).

The complex behaviours of the solutions may result in a more accurate description of the intricate nature of reality, as we have already mentioned in Section 1.

5.1 Space homogeneous case

In this section, we consider Eq. (2.19) and we assume that $\zeta = 1$, in other words, we consider Eq. (2.4). Let

$$F(t,v) = \int_0^t (a * f^\gamma)(s,v)\mathrm{d}s\,,$$

$$\varphi(t,v) = \int_0^t (a * f)(s,v)\, e^{-(\gamma-1)F(s,v)}\mathrm{d}s\,.$$

Then

$$\frac{\partial}{\partial t}F = a * f^\gamma\,, \qquad \frac{\partial}{\partial t}\varphi = a * f\,. \tag{5.1}$$

Equation (2.22) yields

$$f(t,v) = \frac{\overset{\circ}{f}(v)}{e^{F(t,v)}\left(1-(\gamma-1)\overset{\circ}{f}^{\,\gamma-1}(v)\,\varphi(t,v)\right)^{\frac{1}{\gamma-1}}}\,. \tag{5.2}$$

DOI: 10.1201/9781003537595-5

From the definitions of φ and F it follows

$$\overset{\circ}{f}^{\gamma-1}(w)\,\varphi(t,w) = \overset{\circ}{f}^{\gamma-1}(v)\,\varphi(t,v) \quad \text{and} \quad F(t,v) = F(t,w)$$

for any $v, w \in \mathbb{W}$, where the set $\mathbb{W}$ is given by

$$\mathbb{W} = \Big\{ w \in \mathbb{V} : \quad a(w,\widetilde{v}) \leq a(v,\widetilde{v}) \quad \text{and} \\ \overset{\circ}{f}^{\gamma-1}(w)\, a(w,\widetilde{v}) \geq \overset{\circ}{f}^{\gamma-1}(v)\, a(v,\widetilde{v}) \quad \text{for} \quad v,\widetilde{v} \in \mathbb{V} \Big\}. \tag{5.3}$$

Integrating we obtain

$$\frac{\partial}{\partial t} F(t,v) = \int_{\mathbb{V}} \frac{a(v,w)\,\overset{\circ}{f}^{\gamma}(w)}{\exp\Big(\gamma\, F(t,w)\Big)\Big(1-(\gamma-1)\,\overset{\circ}{f}^{\gamma-1}(w)\,\varphi(t,w)\Big)^{\frac{\gamma}{\gamma-1}}}\, dw\,,$$

and then next

$$\frac{\partial}{\partial t}\varphi(t,v) = \frac{(a*f)(t,v)}{\exp\Big((\gamma-1)\,F(t,v)\Big)} =$$

$$= \frac{1}{\exp\Big(-(\gamma-1)\,F(t,v)\Big)} \int_{\mathbb{V}} \frac{a(v,w)\,\overset{\circ}{f}(w)}{\exp\Big(F(t,w)\Big)\Big(1-(\gamma-1)\overset{\circ}{f}^{\gamma-1}(w)\,\varphi(t,w)\Big)^{\frac{1}{\gamma-1}}}\, dw\,.$$

It follows

$$\ln \frac{\partial}{\partial t}\varphi(t,v) = \ln\Big((a*f)(t,v)\Big) - (\gamma-1)F(t,v)\,. \tag{5.4}$$

Noting Eq. (5.2), we observe that a blow–up can arise when the denominator

$$\mathfrak{d}\,(t,v) = 1 - (\gamma-1)\,\overset{\circ}{f}^{\gamma-1}(v)\,\varphi(t,v)$$

tends to 0.

Let $\mathbb{W} \neq \emptyset$. Then the denominator $\mathfrak{d}\,(t,v)$ for $v \in \mathbb{W}$ is the smallest,

$$\mathfrak{d}\,(t,v) > \mathfrak{d}\,(t,w)\,, \qquad v \in \mathbb{V}\backslash\mathbb{W}, \quad w \in \mathbb{W}\,,$$

and $\mathfrak{d}\,(t,v)$ is constant on $\mathbb{W}$.

In fact, the set $\mathbb{W}$ was defined in a way to identify the variables $v \in \mathbb{V}$ that can experience a blow–up, since it is the set of variables where the denominator $\mathfrak{d}\,(t,v)$, referred to $f = f(t,v)$, reaches its smallest value.

Thus, a blow–up can occur only in the set $\mathbb{W}$.

We are using the following auxiliary sets

$$\mathbb{W}_a = \Big\{ w \in \mathbb{V} : \quad a(w,\widetilde{v}) \leq a(v,\widetilde{v}) \quad \text{for} \quad v,\widetilde{v} \in \mathbb{V} \Big\},$$

and

$$\mathbb{W}_0 = \Big\{ w \in \mathbb{W}_a : \quad \overset{\circ}{f}(w) \geq \overset{\circ}{f}(v) \quad \text{for} \quad v \in \mathbb{W}_a \Big\}.$$

We have

$$a(w,\widetilde{v}) \leq a(v,\widetilde{v})\,, \qquad w \in \mathbb{W}, \quad v,\widetilde{v} \in \mathbb{V}.$$

Therefore,

$$\mathbb{W} \subset \mathbb{W}_a ,$$

and

$$a(w, \widetilde{v}) = a(\widetilde{w}, \widetilde{v}) > 0 , \qquad \widetilde{w} \in \mathbb{W}_a .$$

Moreover

$$\overset{\circ}{f}{}^{\gamma-1}(w)\, a(w, \widetilde{v}) \geq \overset{\circ}{f}{}^{\gamma-1}(\widetilde{w})\, a(\widetilde{w}, \widetilde{v}) ,$$

and

$$\overset{\circ}{f}(w) \geq \overset{\circ}{f}(\widetilde{w}) .$$

It follows $\mathbb{W} \subset \mathbb{W}_0$ and we have

$$\mathbb{W} \subset \mathbb{W}_0 \subset \mathbb{W}_a . \tag{5.5}$$

The set $\mathbb{W}$ can be interpreted as a generalization of the analogous set discussed in the preceding chapter. If $w \in \mathbb{W}$, then the function a, viewed as a function of its first variable, achieves its global minimum at w, while $\overset{\circ}{f}$ reaches its global maximum at the same point w. Therefore, a particular condition connecting the interaction rate a and the initial datum $\overset{\circ}{f}$ is required for $|\mathbb{W}| > 0$.

In the following, we present a generalization of Theorem 5 that reveals the global existence for case $|\mathbb{W}| > 0$; see Ref. [137], Theorem 4.1.

Theorem 8 *Let*

- *Assumption 4 be satisfied*
- $\overset{\circ}{f} \in L^\infty$ *is a probability density such that* $|\mathbb{W}| > 0$ *.*

Then for any $t > 0$*, there exists a unique solution* $f = f(t)$ *of Eq.* (2.19)*, with* $\zeta = 1$ *and the initial datum* $\overset{\circ}{f}$*, in* L^∞*. The solution* $f = f(t)$ *is a probability density for each* $t > 0$ *.*

Proof. The proof follows Ref. [137]. The local–in–time existence, uniqueness, positivity, and conservation of the L^1–norm follow immediately; cf. Proposition 1.
Equation (5.5) is satisfied and we have

$$a(w, v) = a(\widetilde{w}, v) , \qquad w, \widetilde{w} \in \mathbb{W}_a , \quad v \in \mathbb{V} .$$

It follows that

$$F(t, w) = F(t, \widetilde{w})$$

and

$$\varphi(t, w) = \varphi(t, \widetilde{w}) .$$

For $w, \widetilde{w} \in \mathbb{W}_0$, by

$$\overset{\circ}{f}(w) = \overset{\circ}{f}(\widetilde{w}) ,$$

it follows

$$f(t, w) = f(t, \widetilde{w}) .$$

The non–negativity of the solution implies

$$|\mathbb{W}_0|\, f(t, w) = \int\limits_{\mathbb{W}_0} f(t, v)\, dv \leq \int\limits_{\mathbb{V}} f(t, v)\, dv = 1 \qquad w \in \mathbb{W}_0 .$$

Due to

$$0 < |\mathbb{W}| \leq |\mathbb{W}_0| ,$$

we obtain the upper bound of the solution in subset $\mathbb{W}$, and therefore the result of global existence (and uniqueness). ■ QED

Next, we can state blow–up properties in a way similar to that of the constant a – see Chapter 4. However, we need some rather restrictive assumptions.

Let a be factorized; see Ref. [137], that is

$$a(v,w) = a_1(v)a_1(w)\,, \qquad \text{for} \quad v,w \in \mathbb{V}\,, \tag{5.6}$$

with some function a_1.

Let

$$\Phi(\varphi,v) = \left(\int_{\mathbb{V}} \frac{\overset{\circ}{f}(w)}{\left(1-(\gamma-1)\overset{\circ}{f}^{\gamma-1}(w)\,\frac{a_1(w)}{a_1(v)}\,\varphi\right)^{\frac{1}{\gamma-1}}}\,dw \right)^{1-\gamma}\,,$$

$$\varphi_0(v) = \frac{a_1(v)}{(\gamma-1)\left\|\overset{\circ}{f}^{\gamma-1}a_1\right\|_\infty}\,, \tag{5.7}$$

and

$$\Phi(\varphi_0,v) = \lim_{\varphi\to(\varphi_0(v))^-} \Phi(\varphi,v)\,.$$

Changing the integral over $\mathbb{V}$ to the integral over $\mathbb{W}_a$ we obtain the corresponding definition of $\Phi_a(\varphi,v)$.

To address technical problems when $|\mathbb{W}| = 0$, we introduce the following helpful lemma before presenting our results on blow–up and global existence; see Lemma 4.1 in Ref. [137]. Taking into account Assumption 4 we introduce

$$\underline{a} = \frac{a_-}{a_+} \in (0,1) \tag{5.8}$$

Lemma 1 *Let*

- *Assumption 4 together with Eq. (5.6) be satisfied,*
- *the probability density* $\overset{\circ}{f}$ *be in* L^∞,
- $|\mathbb{W}| = 0$, *but* $\mathbb{W} \neq \emptyset$.

Then any solution f to Eq. (2.19), with $\zeta = 1$, corresponding to the initial datum $\overset{\circ}{f}$, a priori satisfies

1.

$$\begin{aligned}\left(\underline{a}-\tfrac{1}{\underline{a}}\right)(a*f^\gamma)(t,v) \;&\leq\; \frac{(a*\frac{\partial}{\partial t}f)(t,v)}{(a*f)(t,v)} \;\leq \\ &\leq\; \left(\tfrac{1}{\underline{a}}-\underline{a}\right)(a*f^\gamma)(t,v)\,,\end{aligned} \tag{5.9}$$

for all $v \in \mathbb{V}$;

2.

$$\underline{a}\,F(t,v) \leq F(t,w) \leq \frac{1}{\underline{a}}\,F(t,v)\,, \tag{5.10}$$

for all $v,w \in \mathbb{V}$, and there are constants $\mathfrak{c}_1 > 0$ and $\mathfrak{c}_2 > 0$ such that

$$\mathfrak{c}_1\left(\frac{\partial}{\partial t}\varphi(t,v)\right)^{\frac{1}{\underline{a}\,(\gamma-1)}} \leq \frac{1}{\exp\left(F(t,w)\right)} \leq \mathfrak{c}_2\left(\frac{\partial}{\partial t}\varphi(t,v)\right)^{\frac{\underline{a}}{\gamma-1}}\,, \tag{5.11}$$

for all $v \in \mathbb{V}$;

3. *For each constant* $\mathfrak{c}_0 > 0$ *there exists a constant* $\mathfrak{c}_1 > 0$ *such that*

$$\begin{aligned}
&\left(a * \overset{\circ}{f}\right)^{-\frac{\underline{a}}{\gamma-1}} + \\
&+ \mathfrak{c}_0 \int_{\mathbb{V}} \overset{\circ}{f}(w) \left(\frac{1}{\left(1-(\gamma-1)\overset{\circ}{f}^{\,\gamma-1}(w)\,\varphi(t,w)\right)^{\frac{1}{\gamma-1}}} - 1 \right) \mathrm{d}w \le \\
&\le \mathfrak{c}_1 \int_{\mathbb{V}} \frac{\overset{\circ}{f}(w)}{\left(1-(\gamma-1)\overset{\circ}{f}^{\,\gamma-1}(w)\,\varphi(t,w)\right)^{\frac{1}{\gamma-1}}} \, \mathrm{d}w\,;
\end{aligned} \tag{5.12}$$

4. *If*

$$\int_{\mathbb{W}_a} \overset{\circ}{f}(w)\, \mathrm{d}w > 0$$

then, for each constant $\mathfrak{c}_0 > 0$*, there exists a constant* $\mathfrak{c}_1 > 0$ *such that*

$$\begin{aligned}
&\mathfrak{c}_1 \int_{\mathbb{W}_a} \frac{\overset{\circ}{f}(w)}{\left(1-(\gamma-1)\overset{\circ}{f}^{\,\gamma-1}(w)\,\varphi(t,w)\right)^{\frac{1}{\gamma-1}}} \, \mathrm{d}w \le \\
&\le \left(a * \overset{\circ}{f}\right)^{-\frac{1}{\underline{a}(\gamma-1)}} + \\
&+ \mathfrak{c}_0 \int_{\mathbb{W}_a} \overset{\circ}{f}(w) \left(\frac{1}{\left(1-(\gamma-1)\overset{\circ}{f}^{\,\gamma-1}(w)\,\varphi(t,w)\right)^{\frac{1}{\gamma-1}}} - 1 \right) \mathrm{d}w\,.
\end{aligned} \tag{5.13}$$

Proof. The proof follows Ref. [137]. By Eqs. (2.19), with $\zeta = 1$, and Assumption 4 we have

$$\begin{aligned}
&\left(a * \tfrac{\partial}{\partial t} f\right)(t,v) = \left(a * \left(f^\gamma\,(a*f) - f\,(a * f^\gamma)\right)\right)(t,v) = \\
&= \int_{\mathbb{V}} a(v,w) \Big(f^\gamma(t,w) \int_{\mathbb{V}} a(w,\bar{w}) f(t,\bar{w})\, \mathrm{d}\bar{w} \\
&- f(t,w) \int_{\mathbb{V}} a(w,\bar{w}) f^\gamma(t,\bar{w})\, \mathrm{d}\bar{w} \Big)\, \mathrm{d}w \le \\
&\le \tfrac{1}{\underline{a}} \int_{\mathbb{V}} a(v,w) f^\gamma(t,w) \int_{\mathbb{V}} a(v,\bar{w}) f(t,\bar{w})\, \mathrm{d}\bar{w}\, dw \\
&- \underline{a} \int_{\mathbb{V}} a(v,\bar{w})\, f^\gamma(t,\bar{w}) \int_{\mathbb{V}} a(v,w)\, f(t,w)\, \mathrm{d}w\, \mathrm{d}\bar{w} = \\
&= \left(\tfrac{1}{\underline{a}} - \underline{a}\right) (a * f^\gamma)(t,v)\; (a*f)(t,v)\,.
\end{aligned}$$

Thus the upper estimation in Eq. (5.9) follows. Similarly, we can demonstrate the lower inequality.

To prove Eq. (5.11) we note that by Assumption 4 it follows

$$\underline{a}\,(a*f^{\gamma})(t,v) \le (a*f^{\gamma})(t,w) \le \frac{1}{\underline{a}}\,(a*f^{\gamma})(t,v)\,.$$

Therefore,

$$\underline{a}\,F(t,v) \le F(t,w) \le \frac{1}{\underline{a}}\,F(t,v)\,.$$

By Eq. (5.4) we obtain

$$\frac{1}{\exp\left(F(t,v)\right)} = \frac{\left(\frac{\partial}{\partial t}\varphi(t,v)\right)^{\frac{1}{\gamma-1}}}{\Big((a*f)(t,v)\Big)^{\frac{1}{\gamma-1}}}\,.$$

Taking into account Assumption 4, we obtain uniform estimates

$$a_- \;\le a*f\;\le a_+\,,$$

for any probability density f. Therefore, there exists a positive constant $\mathfrak{c}_2$ such that

$$\frac{1}{\exp(F(t,w))} \le \frac{1}{\exp(\underline{a}\,F(t,v))} = \frac{\left(\frac{\partial}{\partial t}\varphi(t,v)\right)^{\frac{\underline{a}}{\gamma-1}}}{\Big((a*f)(t,v)\Big)^{\frac{\underline{a}}{\gamma-1}}} \le$$

$$\le \quad \mathfrak{c}_2\left(\tfrac{\partial}{\partial t}\varphi(t,v)\right)^{\frac{\underline{a}}{\gamma-1}}\,.$$

Similar arguments leads to

$$\mathfrak{c}_1\left(\frac{\partial}{\partial t}\varphi(t,v)\right)^{\frac{1}{\underline{a}\,(\gamma-1)}} \le \frac{1}{\exp\left(\frac{1}{\underline{a}}\,F(t,v)\right)} \le \frac{1}{\exp\left(F(t,w)\right)}\,,$$

with a positive constant $\mathfrak{c}_1$.

Now we are going to prove Eq. (5.12). We fix $\mathfrak{c}_0 > 0$. Since $a*\mathring{f}$ is bounded and separated from zero, there exists a constant $\mathfrak{c}_1 > 0$ such that $\mathfrak{c}_1 \ge \mathfrak{c}_0$ and

$$1 \le \mathfrak{c}_1\left(a*\mathring{f}\right)^{\frac{\underline{a}}{\gamma-1}}\,.$$

Then we can prove Eq. (5.12) in the following way

$$\frac{1}{\left(a*\mathring{f}\right)^{\frac{\underline{a}}{\gamma-1}}} + \mathfrak{c}_0\int_{\mathbb{V}}\mathring{f}(w)\left(\frac{1}{\left(1-(\gamma-1)\mathring{f}^{\,\gamma-1}(w)\,\varphi(t,w)\right)^{\frac{1}{\gamma-1}}}-1\right)dw \le$$

$$\le \quad \mathfrak{c}_1 + \mathfrak{c}_1\int_{\mathbb{V}}\mathring{f}\,(w)\left(\frac{1}{\left(1-(\gamma-1)\mathring{f}^{\,\gamma-1}(w)\,\varphi(t,w)\right)^{\frac{1}{\gamma-1}}}-1\right)dw \;=$$

$$= \quad \mathfrak{c}_1\int_{\mathbb{V}}\frac{\mathring{f}(w)}{\left(1-(\gamma-1)\mathring{f}^{\,\gamma-1}(w)\,\varphi(t,w)\right)^{\frac{1}{\gamma-1}}}\,dw\,,$$

where the identity $\int\limits_{\mathbb{V}} \overset{\circ}{f}\, \mathrm{d}w = 1$ was used.

Finally, we prove Eq. (5.13). We fix $\mathfrak{c}_0 > 0$. There exists a positive constant $\mathfrak{c}_1$ such that

$$\mathfrak{c}_1 \leq \mathfrak{c}_0 \int\limits_{\mathbb{W}_a} \overset{\circ}{f}(w)\, \mathrm{d}w \,,$$

and

$$\left(a * \overset{\circ}{f}\right)^{\frac{1}{\underline{a}\,(\gamma-1)}} \leq \frac{1}{\mathfrak{c}_1}\,.$$

Therefore, Eq. (5.13) follows. ■ QED

Moreover, we have; see Theorem 4.2 in Ref. [137],

Theorem 9 *Let*

- *Assumption 4 together with Eq. (5.6) be satisfied,*
- *The probability density* $\overset{\circ}{f}$ *be in* L^∞ ,
- $\overset{\circ}{f}$ *together with* a *be continuous functions,*
- $|\mathbb{W}| = 0$, *and* $\mathbb{W} \neq \emptyset$,
- $\varphi_0 = \varphi_0(v)$, $v \in \mathbb{W}$, *is given by Eq.* (5.7),
- *One of the following conditions hold*
  1. $\Phi(\varphi_0, v) > 0$,
  2. $\Phi(\varphi_0, v) = 0$ *and*

$$\int\limits_0^{\varphi_0} \frac{1}{\left(\Phi(\varphi, v)\right)^{\underline{a}}}\, \mathrm{d}\varphi < \infty\,. \tag{5.14}$$

Then the solution $f = f(t)$ *to Eq.* (2.19), *with* $\zeta = 1$, *corresponding to the initial datum* $\overset{\circ}{f}$, *blows up in finite time* $\tau > 0$, *for each* $v \in \mathbb{W}$.

Proof. The proof follows Ref. [137]. We start with the observation that if $\overset{\circ}{f}$ and a are continuous functions, then the corresponding solution is *a priori* continuous. Let $v \in \mathbb{V}$.

Differentiating Eq. (5.4) and using Eq. (5.1), results in

$$\frac{\frac{\partial^2}{\partial t^2}\varphi(t, v)}{\frac{\partial}{\partial t}\varphi(t, v)} = \frac{\left(a * \frac{\partial}{\partial t} f\right)(t, v)}{\left(a * f\right)(t, v)} - (\gamma - 1)\left(a * f^\gamma\right)(t, v)\,. \tag{5.15}$$

Applying Lemma 1, Item (2), we obtain

$$\frac{\frac{\partial^2}{\partial t^2}\varphi(t, v)}{\frac{\partial}{\partial t}\varphi(t, v)} \geq -\mathfrak{c}\left(a * f^\gamma\right)(t, v)\,,$$

where $\mathfrak{c}$ denotes positive constants: Our convention is that when $\mathfrak{c}$ is included in a formula, it means that the formula is true with a constant $\mathfrak{c} > 0$.

By Eq. (5.2) and Lemma 1, Item (2), it follows

$$\frac{\frac{\partial^2}{\partial t^2}\varphi(t,v)}{\frac{\partial}{\partial t}\varphi(t,v)} \geq -\mathfrak{c}\int_{\mathbb{V}} \mathring{f}(w) \frac{(\gamma-1)\mathring{f}^{\gamma-1}(w)\frac{\partial}{\partial t}\varphi(t,w)e^{-F(t,w)}}{\left(1-(\gamma-1)\mathring{f}^{\gamma-1}(w)\,\varphi(t,w)\right)^{\frac{\gamma}{\gamma-1}}}\,\mathrm{d}w \geq$$

$$\geq -\mathfrak{c}\int_{\mathbb{V}} \mathring{f}(w) \frac{(\gamma-1)\mathring{f}^{\gamma-1}(w)\frac{\partial}{\partial t}\varphi(t,w)\left(\frac{\partial}{\partial t}\varphi(t,v)\right)^{\frac{a}{\gamma-1}}}{\left(1-(\gamma-1)\mathring{f}^{\gamma-1}(w)\,\varphi(t,w)\right)^{\frac{\gamma}{\gamma-1}}}\,\mathrm{d}w =$$

$$= -\mathfrak{c}\left(\frac{\partial}{\partial t}\varphi(t,v)\right)^{\frac{a}{\gamma-1}} \int_{\mathbb{V}} \mathring{f}(w)\,\frac{\partial}{\partial t}\left(\frac{1}{\left(1-(\gamma-1)\,\mathring{f}^{\gamma-1}(w)\,\varphi(t,w)\right)^{\frac{1}{\gamma-1}}}\right)\mathrm{d}w\,.$$

Thus,

$$\frac{\partial}{\partial t}\varphi(t,v) \geq \left(\frac{1}{\left(a*\mathring{f}\right)^{\frac{a}{\gamma-1}}} + \right.$$

$$\left. + \ \mathfrak{c}\int_{\mathbb{V}} \mathring{f}(w)\left(\frac{1}{\left(1-(\gamma-1)\mathring{f}^{\gamma-1}(w)\,\varphi(t,w)\right)^{\frac{1}{\gamma-1}}} - 1\right)\mathrm{d}w\right)^{-\underline{a}(\gamma-1)}.$$

Assume that $v \in \mathbb{W}$. By applying Lemma 1, Item (3), it is possible to choose a positive constant, again denoted by $\mathfrak{c}$, so that

$$\frac{\partial}{\partial t}\varphi(t,v) \geq \frac{\mathfrak{c}}{\left(\int_{\mathbb{V}} \frac{\mathring{f}(w)}{\left(1-(\gamma-1)\mathring{f}^{\gamma-1}(w)\,\varphi(t,w)\right)^{\frac{1}{\gamma-1}}}\,\mathrm{d}w\right)^{\underline{a}(\gamma-1)}}.$$

Using Eq. (5.6), by the property of monotonicity, see the definition of $\mathbb{W}$, we have

$$\varphi(t,w) \leq \frac{a_1(w)}{a_1(v)}\,\varphi(t,v) \qquad \text{for} \quad w \in \mathbb{V},\ v \in \mathbb{W},$$

and

$$\frac{\partial}{\partial t}\varphi(t,v) \geq \frac{\mathfrak{c}}{\left(\int_{\mathbb{V}} \frac{\mathring{f}(w)}{\left(1-(\gamma-1)\,\mathring{f}^{\gamma-1}(w)\,\frac{a_1(w)}{a_1(v)}\,\varphi(t,v)\right)^{\frac{1}{\gamma-1}}}\,\mathrm{d}w\right)^{\underline{a}\,(\gamma-1)}},$$

that is

$$\frac{\partial}{\partial t}\varphi(t,v) \geq \mathfrak{c}\left(\Phi(\varphi(t,v),v)\right)^{\underline{a}}.$$

We establish the comparison equation for every $v \in \mathbb{W}$

$$\frac{\mathrm{d}}{\mathrm{d}t}\Xi = \mathfrak{c}\left(\Phi(\Xi,v)\right)^{\underline{a}},$$

The right-hand side does not depend on the selection of $v \in \mathbb{W}$. Two potential scenarios for a blow-up can be recognized: one where the positive stationary solution φ_0 possesses a positive integral, and the other where it is unbounded and Eq. (5.14) is satisfied. This concludes the proof. ■ QED

The continuity assumption on the initial data and function a is quite restrictive and probably unnatural. It should be noted that the statement could be more general and the assumption may be relaxed, e.g. to the *càdlàg* functions in one–dimensional case $\mathfrak{n} = 1$.

Following Theorem 4.3 in Ref. [137] we generalize now the part of Theorem 6 that states the global existence when $|\mathbb{W}| = 0$.

Theorem 10 *Let*

- $$\gamma + \underline{a} > 1 + \frac{1}{\underline{a}},$$
- *Assumption 4 together with Eq. (5.6) be satisfied,*
- *A probability density* $\overset{\circ}{f}$ *is in* L^∞*,*
- $|\mathbb{W}| = 0$*, and* $\mathbb{W} \neq \emptyset$*,* $\int\limits_{\mathbb{W}_a} \overset{\circ}{f} \, \mathrm{d}w > 0$.
- $\overset{\circ}{f}$ *and* a *are continuous functions,*
- $\Phi_a(\varphi_0, v) = 0$*,* $v \in \mathbb{W}$*, and*

$$\int\limits_0^{\varphi_0} \frac{1}{\Big(\Phi_a(\varphi, v)\Big)^{\frac{1}{\underline{a}}}} \, \mathrm{d}\varphi = +\infty. \tag{5.16}$$

Then, for any $\tau > 0$*, in* $C^1(0, \tau; L^\infty)$*, there exists a unique solution* $f = f(t)$ *of Eq. (2.19), with* $\zeta = 1$*, corresponding to the initial datum* $\overset{\circ}{f}$*. Moreover, the solution* $f(t)$ *is a probability density and is continuous for* $t \geq 0$.

Proof. We follow Ref. [137]. The scenario with the most unfavorable outcome in Eq. (5.2) occurs when the variable v is an element of the set $\mathbb{W}$. With this in mind, estimates will be carried out in this case. Differentiating Eq. (5.4) and using Eq. (5.1), yields Eq. (5.15). Applying Lemma 1 (1) we have

$$\frac{\frac{\partial^2}{\partial t^2}\varphi(t,v)}{\frac{\partial}{\partial t}\varphi(t,v)} =$$

$$= \frac{(a * \frac{\partial}{\partial t} f)(t,v)}{(a * f)(t,v)} - (\gamma - 1)\,(a * f^\gamma)(t,v) \leq$$

$$\leq -\Big(\gamma + \underline{a} - 1 - \tfrac{1}{\underline{a}}\Big)(a * f^\gamma)(t,v) =$$

$$= -\Big(\gamma + \underline{a} - 1 - \tfrac{1}{\underline{a}}\Big) \int\limits_{\mathbb{V}} \frac{a(v,w)\, \overset{\circ}{f}{}^{\gamma}(w)}{\exp(\gamma F(t,w)) \Big(1 - (\gamma-1)\overset{\circ}{f}{}^{\gamma-1}(w)\, \varphi(t,w)\Big)^{\frac{\gamma}{\gamma-1}}} \, \mathrm{d}w\,.$$

Taking into account $|\mathbb{W}_a| > 0$ for $v \in \mathbb{W}$ it follows

$$\frac{\frac{\partial^2}{\partial t^2}\varphi(t,v)}{\frac{\partial}{\partial t}\varphi(t,v)} \leq -\mathfrak{c} \int\limits_{\mathbb{W}_a} \frac{\overset{\circ}{f}{}^{\gamma}(w)\, \exp\left((1-\gamma)\, F(t,w)\right)}{\exp\left(F(t,w)\right) \left(1-(\gamma-1)\, \overset{\circ}{f}{}^{\gamma-1}(w)\, \varphi(t,w)\right)^{\frac{\gamma}{\gamma-1}}}\, dw\,,$$

where again $\mathfrak{c}$ is a (generic) positive constant.

By Lemma 1, Item (2),

$$\frac{1}{\exp((\gamma-1)\, F(t,w))\, \exp(F(t,w))} = \frac{\frac{\partial}{\partial t}\varphi(t,w)}{(a*f(t,w))\, \exp(F(t,w))} \geq$$

$$\geq \mathfrak{c}\, \frac{\partial}{\partial t}\varphi(t,w) \Big(\frac{\partial}{\partial t}\varphi(t,v)\Big)^{\frac{1}{\underline{a}(\gamma-1)}}.$$

Thus, for $v \in \mathbb{W}$,

$$\frac{\frac{\partial^2}{\partial t^2}\varphi(t,v)}{\frac{\partial}{\partial t}\varphi(t,v)} \left(\frac{\partial}{\partial t}\varphi(t,v)\right)^{-\frac{1}{\underline{a}\,(\gamma-1)}} \leq$$

$$\leq -\mathfrak{c} \int\limits_{\mathbb{W}_a} \frac{\overset{\circ}{f}(w)\, (\gamma-1)\, \overset{\circ}{f}{}^{\gamma-1}(w)\, \frac{\partial}{\partial t}\varphi(t,w)}{\left(1-(\gamma-1)\, \overset{\circ}{f}{}^{\gamma-1}(w)\, \varphi(t,w)\right)^{\frac{\gamma}{\gamma-1}}}\, dw =$$

$$= -\mathfrak{c} \int\limits_{\mathbb{W}_a} \overset{\circ}{f}(w)\, \frac{\partial}{\partial t} \left(\frac{1}{\left(1-(\gamma-1)\overset{\circ}{f}{}^{\gamma-1}(w)\, \varphi(t,w)\right)^{\frac{1}{\gamma-1}}}\right) dw\,.$$

Taking the integral results in

$$\left(\frac{\partial}{\partial t}\varphi(t,v)\right)^{-\frac{1}{\underline{a}\,(\gamma-1)}} \geq \left(a * \overset{\circ}{f}\right)^{-\frac{1}{\underline{a}\,(\gamma-1)}}(v) +$$

$$+ \mathfrak{c} \int\limits_{\mathbb{W}_a} \overset{\circ}{f}(w) \left(\frac{1}{\left(1-(\gamma-1)\, \overset{\circ}{f}{}^{\gamma-1}(w)\, \varphi(t,w)\right)^{\frac{1}{\gamma-1}}} - 1\right) dw$$

and, by Lemma 1 Item (4), we obtain

$$\frac{\partial}{\partial t}\varphi(t,v) \leq \Bigg((a*\overset{\circ}{f})^{-\frac{1}{\underline{a}(\gamma-1)}}(v) +$$

$$+ \mathfrak{c} \int\limits_{\mathbb{W}_a} \overset{\circ}{f}(w) \Bigg(\frac{1}{\left(1-(\gamma-1)\overset{\circ}{f}{}^{\gamma-1}(w)\, \frac{a_1(w)}{a_1(v)}\, \varphi(t,v)\right)^{\frac{1}{\gamma-1}}} - 1\Bigg)\, dw\Bigg)^{-\frac{1}{\underline{a}\,(\gamma-1)}} \leq$$

$$\leq \mathfrak{c} \left(\int\limits_{\mathbb{W}_a} \frac{\overset{\circ}{f}(w)}{\left(1-(\gamma-1)\overset{\circ}{f}{}^{\gamma-1}(w)\, \frac{a_1(w)}{a_1(v)}\, \varphi(t,v)\right)^{\frac{1}{\gamma-1}}}\, dw\right)^{-\frac{1}{\underline{a}\,(\gamma-1)}} =$$

$$= \mathfrak{c} \left(\Phi_a(\varphi(t,v), v)\right)^{\frac{1}{\underline{a}}}.$$

We now define the comparison equation as follows

$$\frac{\mathrm{d}}{\mathrm{d}t}\Xi(t,v) = \mathfrak{c}\left(\Phi_a(\Xi,v)\right)^{\frac{1}{a}}.$$

The right–hand side is independent of the selection of the variable $v \in \mathbb{W}$. The global existence of Ξ follows by the assumption

$$\Phi_a(\varphi_0, v) = 0\,,$$

together with Eq. (5.16). Thus $\varphi(t,v)$ exists globally which ends the proof. ■ QED

Referring to Theorem 10, as before we can observe that the continuity assumption on the initial data and function a is quite restrictive and can be relaxed.

5.2 Space inhomogeneous case

In this section, still following Ref. [137], we consider the space–dependent case; cf. Section 2.4, see Eq. (2.26), which is a generalization of Eq. (4.9) in Chapter 4.

We can use the methods developed in the previous Section 5.1 to deal with Eq. (2.26), i.e.

$$\begin{aligned}\frac{\partial}{\partial t}f(t,x,v) + v\cdot\frac{\partial}{\partial x}f(t,x,v) &= \\ = \; (a*f)(t,v)f^{\gamma}(t,x,v) &- (a*f^{\gamma})(t,v)f(t,x,v)\end{aligned} \tag{5.17}$$

with the initial data

$$f(0,x,v) = \overset{\circ}{f}(x,v)\,,$$

that is assumed to be a probability density on $\mathbb{D}\times\mathbb{V}$ and $\mathbb{D} = \mathbb{R}^{\mathfrak{n}}$. Similarly, we can move forward with $\mathbb{T}^{\mathfrak{n}}$, the $\mathfrak{n}$–dimensional torus corresponding to the periodic boundary conditions, or a general domain with the Kaniel–Shinbrot boundary conditions; see Ref. [116]. The methods are exactly as for Eq. (4.9) in Section 4.

We analyse the equation by examining it along the characteristics

$$\frac{\partial}{\partial t}f^{\#}(t,x,v) = (a*f)(t,v)(f^{\gamma})^{\#}(t,x,v) - f^{\#}(t,x,v)(a*f^{\gamma})(t,v)\,. \tag{5.18}$$

The integration results in

$$f^{\#}(t,x,v) = \frac{\overset{\circ}{f}(x,v)}{\exp\left(F(t,v)\right)\left(1-(\gamma-1)\overset{\circ}{f}(x,v)\varphi(t,v)\right)^{\frac{1}{\gamma-1}}}\,,$$

where

$$F(t,v) = \int_0^t (a*f^{\gamma})(s,v)\,\mathrm{d}s$$

and

$$\varphi(t,v) = \int_0^t \frac{(a*f)(s,v)}{\exp\left((\gamma-1)F(s,v)\right)}\,\mathrm{d}s\,.$$

We can evidently demonstrate theorems analogous to those in Section 5.1 in the present context.

5.3 Further analysis

In this part, following Ref. [137], we examine the problem of blow–up compared to boundedness of L^p-norm of solutions, where $1 \leq p < \infty$, over finite time intervals. It is evident that the global existence in L^∞ guarantees the boundedness of any L^p-norm for $p \geq 1$ over compact time intervals. However, it is not clear whether the blow–ups studied in the previous section result in blow–ups in the L^γ–norm.

The objective is to define blow-up contrasted with global boundedness in L^γ utilizing the subsequent quantities; cf. Ref. [137],

$$\overset{\circ}{M}_p = \int_{\mathbb{V}} \overset{\circ}{f}^{\,p}(v)\,\mathrm{d}v\,, \qquad M_p(t) = \int_{\mathbb{V}} f^p(t,v)\,\mathrm{d}v\,,$$

related to the initial datum $\overset{\circ}{f}$ and the solution $f = f(t)$, respectively. It follows

$$\begin{aligned}\frac{\mathrm{d}}{\mathrm{d}t} M_p(t) &= p\int_{\mathbb{V}} f^{p-1}(t,v)\,\frac{\partial}{\partial t} f(t,v)\,\mathrm{d}v = \\ &= \frac{p}{2}\Bigg(\int_{\mathbb{V}} f^{p-1+\gamma}(t,v)\int_{\mathbb{V}} a(v,w)\,f(t,w)\,\mathrm{d}w\,\mathrm{d}v + \\ &+ \int_{\mathbb{V}} f^{p-1+\gamma}(t,w)\int_{\mathbb{V}} a(v,w)\,f(t,v)\,\mathrm{d}v\,\mathrm{d}w \\ &- \int_{\mathbb{V}} f^{p}(t,v)\int_{\mathbb{V}} a(v,w)\,f^{\gamma}(t,w)\,\mathrm{d}w\mathrm{d}v \\ &- \int_{\mathbb{V}} f^{p}(t,w)\int_{\mathbb{V}} a(v,w)\,f^{\gamma}(t,v)\,\mathrm{d}v\,\mathrm{d}w\Bigg).\end{aligned}$$

We have

$$\begin{aligned}\frac{\mathrm{d}}{\mathrm{d}t} M_p(t) = \frac{p}{2}\int_{\mathbb{V}^2} a(v,w)\Bigg(& f^{p+\gamma-1}(t,v)\,f(t,w) + f^{p+\gamma-1}(t,w)\,f(t,v) \\ - & f^{p}(t,v)\,f^{\gamma}(t,w) - f^{p}(t,w)\,f^{\gamma}(t,v)\Bigg)\,\mathrm{d}v\,\mathrm{d}w\end{aligned} \tag{5.19}$$

and

$$\begin{aligned}\frac{\mathrm{d}}{\mathrm{d}t} M_p(t) = \frac{p}{2}\int_{\mathbb{V}^2} a(w,v)\,f(t,w)\,f(t,v)\Big(f^{p-1}(t,w) - f^{p-1}(t,v)\Big) \times \\ \times \Big(f^{\gamma-1}(t,w) - f^{\gamma-1}(t,v)\Big)\,\mathrm{d}w\,\mathrm{d}v\,.\end{aligned}$$

We can observe that the quantity M_p, for each $p > 1$, increases. We consider

$$H_\lambda(r) = 1 + \sum_{k=1}^{\infty} \gamma\,(2\gamma-1)\ldots(k\gamma-k+1)\,\overset{\circ}{M}_{k\gamma-k+1}\,\frac{\lambda^k\,r^k}{k!}\,. \tag{5.20}$$

The series given by Eq. (5.20) converges when λr is small, and its convergence radius can be determined by the d'Alembert criterion. We obtain

$$\lim_{k\to\infty} \frac{((k+1)\gamma - k)\ M_{(k+1)\gamma-k,0}}{(k+1)\ M_{k\gamma-k+1,0}}\ \lambda r \ = \ (\gamma-1)\,\lambda r \| \overset{\circ}{f} \|_{\infty}^{\gamma-1},$$

and therefore we should have

$$r < r_0 = \frac{1}{(\gamma-1)\lambda \| \overset{\circ}{f} \|_{\infty}^{\gamma-1}} \tag{5.21}$$

to ensure the convergence.

The global boundedness criterion stated in the following does not require isolating the kernel a from 0. The following theorem refers to the global boundedness in the L^{γ}–norm, cf. Theorem 5.1 in Ref. [137].

Theorem 11 *Let*

- *The probability density* $\overset{\circ}{f}$ *be in* L^{∞},
- $a \in L^{\infty}(\mathbb{V}^2)$,
- *The initial value problem*

$$\frac{\mathrm{d}}{\mathrm{d}t}\,\Xi = \frac{1}{\left(H_{\|a\|_{\infty}}(\Xi)\right)^{\gamma-1}}, \qquad \Xi(0) = 0, \tag{5.22}$$

has a global solution.

Then the solution $f = f(t)$ *of Eq. (2.19) with* $\zeta = 1$, *corresponding to* $\overset{\circ}{f}$, *is a priori bounded in* L^{γ} *in each compact interval of time.*

Proof. Based on the approach in Ref. [137], we initially note that the assumptions of Theorem 11 lead to the conclusion that the solution $\Xi = \Xi(t)$ of Eq. (5.22) is such that

$$\Xi(t) < \frac{1}{(\gamma-1)\,\|a\|_{\infty}\,\| \overset{\circ}{f} \|_{\infty}^{\gamma-1}} \qquad \forall\, t \in (0,\infty), \tag{5.23}$$

where the expression on the right–hand side of Inequality (5.23) defines the radius of convergence of the series (5.20) for $\lambda = \|a\|_{\infty}$; cf. Eq. (5.21).

Taking into account Eq. (5.19) we have

$$\frac{\mathrm{d}}{\mathrm{d}t}\,M_p \le p\,\|a\|_{\infty}\Big(M_{p+\gamma-1} - M_p M_{\gamma}\Big).$$

Therefore

$$\frac{\mathrm{d}}{\mathrm{d}t}\left(M_p(t)\exp\left(p\,\|a\|_{\infty}\int_0^t M_{\gamma}(s)\,\mathrm{d}s\right)\right) \le$$

$$\le \ p\,\|a\|_{\infty} M_{p+\gamma-1}(t)\exp\left(p\,\|a\|_{\infty}\int_0^t M_{\gamma}(s)\,\mathrm{d}s\right),$$

and denoting

$$\widetilde{M}_p(t) = M_p(t) \exp\left(p\, \|a\|_\infty \int_0^t M_\gamma(s)\, \mathrm{d}s \right),$$

$$\beta = \beta(t) = \exp\left((\gamma - 1)\, \|a\|_\infty \int_0^t M_\gamma\, \mathrm{d}\tilde{s} \right),$$

yields

$$\widetilde{M}_p(t) \le \overset{\circ}{M}_p + p\, \|a\|_\infty \int_0^t \frac{\widetilde{M}_{p+\gamma-1}(s)}{\beta(s)}\, \mathrm{d}s\,.$$

Iterating we obtain

$$\widetilde{M}_\gamma(t) \le \overset{\circ}{M}_\gamma + \sum_{k=1}^{\infty} \gamma(2\gamma - 1) \dots (k\gamma - k + 1)\, \|a\|_\infty^k\, \overset{\circ}{M}_{k\gamma-k+1} \times$$

$$\times \int_0^t \frac{1}{\beta(\tau_1)} \int_0^{\tau_1} \frac{1}{\beta(\tau_2)} \cdots \int_0^{t_{k-1}} \frac{1}{\beta(t_k)}\, \mathrm{d}\tau_k \dots \mathrm{d}\tau_1\,,$$

and

$$M_\gamma(t) \exp\left(\|a\|_\infty \int_0^t M_\gamma(\tilde{s}) \mathrm{d}\tilde{s} \right) \le$$

$$\le \frac{\overset{\circ}{M}_\gamma}{\beta(t)} + \sum_{k=1}^{\infty} \gamma(2\gamma - 1) \dots (k\gamma - k + 1) \times$$

$$\times \|a\|_\infty^k \frac{\overset{\circ}{M}_{k\gamma-k+1}}{\beta(t)} \int_0^t \frac{1}{\beta(\tau_1)} \int_0^{\tau_1} \frac{1}{\beta(\tau_2)} \cdots \int_0^{t_{k-1}} \frac{1}{\beta(\tau_k)}\, \mathrm{d}\tau_k \dots \mathrm{d}\tau_1\,.$$

Performing integration over the interval $[0, t]$ results in

$$\frac{\beta^{\frac{1}{\gamma-1}}(t) - 1}{\|a\|_\infty} \le \int_0^t \frac{\overset{\circ}{M}_\gamma}{\beta(\tau_1)} + \sum_{k=1}^{\infty} \gamma(2\gamma - 1) \dots (k\gamma - k + 1) \times$$

$$\times \|a\|_\infty^k \frac{\overset{\circ}{M}_{k\gamma-k+1}}{\beta(t)} \int_0^{\tau_1} \frac{1}{\beta(\tau_2)} \int_0^{\tau_2} \frac{1}{\beta(\tau_3)} \cdots \int_0^{\tau_k} \frac{1}{\beta(t_{k+1})}\, \mathrm{d}t_{k+1} \dots \mathrm{d}\tau_1\,.$$

and then

$$\beta(t) \le \left(H_{\|a\|_\infty} \left(\int_0^t \frac{1}{\beta(s)}\, \mathrm{d}s \right) \right)^{\gamma-1}. \tag{5.24}$$

The quantities M_p are increasing for $p > 1$, as we already note and therefore β is also an increasing function. The greatest among the increasing functions satisfying Inequality (5.24) is the one for which the inequality transforms into equality. Thus,

$$\beta(t) \le \frac{1}{\frac{\mathrm{d}}{\mathrm{d}t}\, \Xi(t)} = \left(H_{\|a\|_\infty} (\Xi(t)) \right)^{\gamma-1}, \qquad t > 0\,,$$

where Ξ is the solution to Eq. (5.22).

Keeping in mind that Ξ is strictly increasing and Eq. (5.23) holds we see that $\beta(t)$ is bounded on any compact interval, and thus

$$M_\gamma(t) = \|f(t, .)\|_\gamma^\gamma$$

is also bounded on any compact time interval. This ends the proof. ■ QED

The following theorem refers to the blow–up of the L^γ–norm; cf. Theorem 5.2 in Ref. [137].

Theorem 12 *Let*

- *The probability density* $\overset{\circ}{f}$ *be in* L^∞,
- *Assumption 4 be satisfied,*
- *The Cauchy problem*

$$\frac{\mathrm{d}}{\mathrm{d}t}\Xi = \frac{1}{\left(H_{a_-}(\Xi)\right)^{\gamma-1}}, \qquad \Xi(0) = 0, \tag{5.25}$$

where a_- *is given in Assumption 4, have a solution only for* $t \in [0, \tau)$, *for some finite* $\tau > 0$.

Then the L^γ*–norm of the solution* $f = f(t)$ *of Eq.* (2.19), *with* $\zeta = 1$, *blows up before* τ.

Proof. Following Ref. [137] we note that by the assumptions of Theorem 12 the solution $\Xi = \Xi(t)$ of Eq. (5.25) is such that

$$\Xi(t) < \frac{1}{(\gamma-1)\, a_- \, \|\overset{\circ}{f}\|_\infty^{\gamma-1}}, \qquad \forall\, t \in [0, \tau), \tag{5.26}$$

and

$$\lim_{t\to\tau} \Xi(t) = \frac{1}{(\gamma-1)\, a_- \, \|\overset{\circ}{f}\|_\infty^{\gamma-1}}. \tag{5.27}$$

By Eq. (5.19) we have

$$\frac{\mathrm{d}}{\mathrm{d}t} M_p \geq p\, a_- \left(M_{p+\gamma-1} - M_p M_\gamma\right),$$

for $p \geq \gamma$. Using iteration, the opportune lower estimate in terms of $\Xi(t)$ is obtained. We have

$$\beta(t) \geq \left(H_{a_-}\left(\int_0^t \frac{1}{\beta(s)}\,\mathrm{d}s\right)\right)^{\gamma-1},$$

and hence

$$\beta(t) \geq \frac{1}{\frac{\mathrm{d}}{\mathrm{d}t}\Xi(t)} = \left(H_{a_-}(\Xi(t))\right)^{\gamma-1}, \qquad t > 0.$$

Taking into account that Ξ is (strictly) increasing and that Eqs. (5.26) and (5.27) are fulfilled, we deduce that $\beta(t)$ is bounded on compact intervals contained in the interval $[0, \tau)$. This ends the proof. ■ QED

Then we have two following statements; see Corollary 5.1 in Ref. [137].

Corollary 3 *Let*

- *The probability density* $\overset{\circ}{f}$ *be in* L^{∞},
- $a \in L^{\infty}(\mathbb{V}^2)$ *but it is not necessarily separated below from zero,*
- *Additionally,*

$$\int_{0}^{\frac{\varphi_0}{\|a\|_\infty}} \left(H_{\|a\|_\infty}(y)\right)^{\gamma-1} \mathrm{d}y = \infty .$$

Then the solution $f = f(t)$ *of Eq.* (2.19) *with* $\zeta = 1$, *corresponding to* $\overset{\circ}{f}$, *is a priori bounded in* L^{γ} *in each compact interval.*

Corollary 4 *Let*

- *The probability density* $\overset{\circ}{f}$ *be in* L^{∞},
- $a \in L^{\infty}(\mathbb{V}^2)$ *be separated from zero, i.e., Assumption 4 be satisfied,*
- *Additionally,*

$$\int_{0}^{\frac{\varphi_0}{a_-}} \left(H_{a_-}(y)\right)^{\gamma-1} \mathrm{d}y < +\infty .$$

Then the L^{γ}*–norm blows up in finite time.*

Remark 1 *By the classical theory of ODE's, the upper bound of the time* τ *of blow–up is defined by*

$$\int_{0}^{\frac{\varphi_0}{a_-}} \left(H_{a_-}(y)\right)^{\gamma-1} \mathrm{d}y .$$

The preceding chapter shows that the rather simple equation (2.19), when $\zeta = 1$, results in a highly intricate behaviour of its solutions. It is evident that this applies equally to both Eqs. (2.23) and (2.24). For the constant kernel a, that is, Eq. (4.1), discussed in Chapter 4, we managed to provide a full explanation of the behaviour of solutions including global existence versus blow-ups. The results completely answer the question of possible creation of swarms giving precise conditions for the blows–up, which can be related to swarming behaviour. In contrast, the kernel a that is not constant, studied in this chapter, is certainly more complex, and we are still far from providing a complete description for the general a and initial data. This is an interesting issue that merits further study.

An interesting situation from both a theoretical and a practical perspective is when the kernel a fulfils, cf. Assumption 6,

$$a(v,w) = \hat{a}(|w-v|) , \tag{5.28}$$

with some function

$$\hat{a} : [0, \infty) \to [0, \infty)$$

that is decreasing. It occurs when agents with similar states have a higher chance of interaction, and each agent can have certain restrictions on changing its state. The theory presented in this chapter relies on Assumption 4 or the factorization in Eq. (5.6), does not cover the scenario described by Eq. (5.28). The following chapter will address certain aspects of short–range interactions.

One can apply methods discussed in the present chapter, like the L^{γ}–blow–up or global existence criteria, to Eq. (2.23) and Eq. (2.24); see Ref. [137]. The quantities $\overset{\circ}{M}_p$ and M_p can be redefined, cf. Ref. [137],

$$\overset{\circ}{M}_p = \int_{\mathbb{R}^n} \int_{\mathbb{V}} \overset{\circ}{f}{}^{p}(x, v)\, \mathrm{d}v\, dx\,, \qquad M_p(t) = \int_{\mathbb{R}^n} \int_{\mathbb{V}} f^p(t, x, v)\, \mathrm{d}v\, \mathrm{d}x\,.$$

and we can apply the equality

$$\int_{\mathbb{R}^n} \frac{\partial}{\partial x} f^p(t, x, v)\, \mathrm{d}x = 0\,.$$

Currently, the unresolved mathematical problems mentioned earlier in this case remain. These are theoretical questions associated with the ability to explain the collective behaviour of complex systems.

Chapter 6

Small range of interactions

In this chapter, we follow Ref. [139] and still refer to $\zeta = 1$ in Eq. (2.19) that is we consider Eq. (2.4).

Chapter 4 dealt with the constant rate $a = a(v, w) \equiv \text{const}$. We were able to provide a complete mathematical description of the possible behaviour of the solutions. However, the case of the constant rate is quite restrictive in terms of potential applications, as we have already discussed. In fact, it would be expected that the kernel a would change with its variables. For example, it would be quite natural that the kernel depends on the distance $|w - v|$ between states v and w. It can, for example, express the fact that interactions are possible when the distance between the states is sufficiently small. Interpreting the parameter v, for example, as an opinion or a position of an agent, such an assumption is quite natural and justified.

This case can be referred to as small–range interactions. In some applications, it is quite natural that the kernel is a symmetric function, which means that if v is replaced with w, the kernel remains the same, i.e. Eq. (2.7) is satisfied. We should note that there are also situations in which the kernel is asymmetric; cf. Section 7.2, Chapter 11, Sections 12.3 and 13.2.3.

We assume; cf. Refs. [134, 138, 139], that the interaction rate a is concentrated close to the diagonal

$$\Big\{(v, w) \in \mathbb{V}^2 \, : \, v = w\Big\},$$

and the parameter $\varepsilon > 0$, tending to 0, describes the "shrinking" effect of the interaction range. This situation, with the appropriate rescaling of a, is used in Chapter 7.

Assumption 6 *Given $\varepsilon > 0$. The function a satisfies*

$$a(v, w) = \hat{a}(|v - w|)\,, \tag{6.1}$$

where $\hat{a} = \hat{a}(x)$ is a non–zero measurable nonnegative function $\mathbb{R}_+ \to \mathbb{R}_+$ such that

$$\hat{a}(x) = 0 \qquad \forall\, x > \varepsilon\,. \tag{6.2}$$

The positive number ε determines the distances at which interactions cannot take place: If the distance between states is greater than ε, then there are no interactions. This is a quite realistic assumption in various particular applications, when only agents with close states can interact. References [136, 137, 170] do not cover the situation of Assumption 6.

In Ref. [139], it was shown both analytically and numerically that in certain situations the emergence of bipolarity can be observed when the range of interactions is small. When the initial data are unimodal, that is, they have a single peak and the peak itself is not sharp, the resulting solution turns bimodal, characterized by the presence of two distinct peaks. Then the solution eventually blows up into two separate points. In contrast, if the initial datum exhibits one sufficiently sharp peak, the solution is characterized by a single peak before experiencing a blow–up.

DOI: 10.1201/9781003537595-6

Despite being proved for particular initial data, the equation offers interesting potential applications. However, the behaviour in the general case is still an open problem.

In the field of political science, the emergence of bimodal distributions may be associated with the formation of two distinct factions within a central political party; see, e.g., Refs. [113, 181, 211].

The conditions guaranteeing bimodality or uni–modality are described in terms of values of the initial datum and its second derivative; see Ref. [139].

Given $\varepsilon > 0$. For simplicity, we consider $\gamma = 2$, $\mathfrak{n} = 1$ and

$$a(v, w) = \chi(|v - w| < \varepsilon), \tag{6.3}$$

where χ is given by Eq. (3.1), that is a special case of Assumption 6.

Then Eq. (2.4) reads

$$\frac{\partial f}{\partial t}(t, v) = f^2(t, v) \int_{v-\varepsilon}^{v+\varepsilon} f(t, w)\, \mathrm{d}w - f(t, v) \int_{v-\varepsilon}^{v+\varepsilon} f^2(t, w)\, \mathrm{d}w. \tag{6.4}$$

The next result refers to the creation of a bimodal distribution versus the preservation of the unimodal distribution before a blow–up; see Theorem 3.1 in Ref. [139].

Theorem 13 *Let*

- $\varepsilon > 0$
- *the initial datum* $\overset{\circ}{f} \in L^1(\mathbb{R}) \cap L^\infty(\mathbb{R})$ *be symmetric*

$$\overset{\circ}{f}(-v) = \overset{\circ}{f}(v), \qquad \forall\, v \in \mathbb{R},$$

non decreasing on $(-\infty, 0)$, *continuous on* $\mathbb{R}$, *and sufficiently smooth on* $[-2\varepsilon, 2\varepsilon]$,

- $\tau > 0$ *define the time interval* $[0, \tau)$ *of the existence of a unique solution of Eq.* (2.4).

Then we have the following two statements.

1. ***If***

$$\overset{\circ}{f}(0) < 2\, \overset{\circ}{f}(\varepsilon), \qquad \frac{\mathrm{d}^2 \overset{\circ}{f}}{\mathrm{d}v^2}(0) = 0, \tag{6.5}$$

and

$$\frac{\mathrm{d}^2 \overset{\circ}{f}}{\mathrm{d}v^2}(v) < 0 \qquad \forall\, v \in [-2\varepsilon, 2\varepsilon] \setminus \{0\}, \tag{6.6}$$

then *there exist* $\bar{\tau} \in (0, \tau)$, $\bar{v} = \bar{v}(t) \in (0, 2\varepsilon]$ *such that* $\bar{v}$ *is an increasing function of* $t \in (0, \bar{\tau})$ *and*

$$\max_{v \in \mathbb{R}} f(t, v) = f(t, \bar{v}(t)) = f(t, -\bar{v}(t)) > f(t, 0), \tag{6.7}$$

for any $t \in (0, \bar{\tau})$, *where* $f = f(t)$ *is the unique solution of Eq.* (2.4) *in* $L^\infty(\mathbb{R})$.

2. *If*

$$\overset{\circ}{f}(0) > 2\, \overset{\circ}{f}(\varepsilon), \tag{6.8}$$

and one of the following conditions is satisfied

(a)

$$\frac{\mathrm{d}^2 \overset{\circ}{f}}{\mathrm{d}\,v^2}(0) < 0\,, \tag{6.9}$$

or

(b)

$$\frac{\mathrm{d}^2 \overset{\circ}{f}}{\mathrm{d}\,v^2}(0) = 0\,, \qquad \frac{\mathrm{d}^2 \overset{\circ}{f}}{\mathrm{d}\,v^2}(v) < 0\,, \qquad \forall\, v \in [-2\varepsilon\,,\, 2\varepsilon] \setminus \{0\}\,, \tag{6.10}$$

then

$$\max_{v\in\mathbb{R}} f\,(t,v) = f(t,0)\,, \tag{6.11}$$

for any $t \in (0,\tau)$, where $f = f(t)$ is a unique solution of Eq. (2.4) in $L^\infty(\mathbb{R})$.

Proof. We follow Ref. [139] – Proof of Theorem 3.1. It is evident that the solution $f = f(t,v)$ maintains the smoothness and symmetry of the initial datum $\overset{\circ}{f} = \overset{\circ}{f}\,(v)$.

The derivative $\frac{\partial f}{\partial v}$ satisfies the following equation

$$\frac{\partial \frac{\partial f}{\partial v}}{\partial t}(t,v) = \frac{\partial f}{\partial v}(t,v)\left(2f(t,v)\int_{v-\varepsilon}^{v+\varepsilon} f(t,w)\,\mathrm{d}w - \int_{v-\varepsilon}^{v+\varepsilon} f^2(t,w)\,\mathrm{d}w\right) +$$

$$+ \quad f^2(t,v)\Big(f(t,v+\varepsilon) - f(t,v-\varepsilon)\Big)$$

$$- \quad f(t,v)\Big(f^2(t,v+\varepsilon) - f^2(t,v-\varepsilon)\Big)\,.$$

By the symmetry of the function $f = f(t,\cdot)$, we have

$$\frac{\partial f}{\partial v}(t,0) = 0\,, \qquad t \in [0,\tau)\,. \tag{6.12}$$

The second derivative $\frac{\partial^2 f}{\partial v^2}$ satisfies the following equation

$$\frac{\partial}{\partial t}\frac{\partial^2 f}{\partial v^2}(t,v) =$$

$$= \quad \frac{\partial^2 f}{\partial v^2}(t,v)\left(2f(t,v)\int_{v-\varepsilon}^{v+\varepsilon} f(t,w)\,\mathrm{d}w - \int_{v-\varepsilon}^{v+\varepsilon} f^2(t,w)\,\mathrm{d}w\right) +$$

$$+ \quad \frac{\partial f}{\partial v}(t,v)\left(2f(t,v)\Big(f(t,v+\varepsilon) - f(t,v-\varepsilon)\Big) + \right.$$

$$\left. + \quad 2\frac{\partial f}{\partial v}(t,v)\int_{v-\varepsilon}^{v+\varepsilon} f(t,w)\,\mathrm{d}w - \Big(f^2(t,v+\varepsilon) - f^2(t,v-\varepsilon)\Big)\right) +$$

$$+ \quad 2f(t,v)\frac{\partial f}{\partial v}(t,v)\Big(f(t,v+\varepsilon) - f(t,v-\varepsilon)\Big)$$

$$- \quad \frac{\partial f}{\partial v}(t,v)\Big(f^2(t,v+\varepsilon) - f^2(t,v-\varepsilon)\Big) +$$

$$+ \quad f^2(t,v)\Big(\frac{\partial f}{\partial v}(t,v+\varepsilon) - \frac{\partial f}{\partial v}(t,v-\varepsilon)\Big)$$

$$- \quad 2f(t,v)\Big(f(t,v+\varepsilon)\frac{\partial f}{\partial v}(t,v+\varepsilon) - f(t,v-\varepsilon)\frac{\partial f}{\partial v}(t,v-\varepsilon)\Big)\,.$$

Putting $v = 0$, we have

$$
\begin{aligned}
&\tfrac{\partial}{\partial t}\tfrac{\partial^2 f}{\partial v^2}(t,0) \;=\\
&= \tfrac{\partial^2 f}{\partial v^2}(t,0)\left(2f(t,0)\int\limits_{-\varepsilon}^{\varepsilon} f(t,w)\,\mathrm{d}w - \int\limits_{-\varepsilon}^{\varepsilon} f^2(t,w)\,\mathrm{d}w\right) +\\
&+ f^2(t,0)\left(\tfrac{\partial f}{\partial v}(t,\varepsilon) - \tfrac{\partial f}{\partial v}(t,-\varepsilon)\right)\\
&- 2f(t,0)\left(f(t,\varepsilon)\tfrac{\partial f}{\partial v}(t,\varepsilon) - f(t,-\varepsilon)\tfrac{\partial f}{\partial v}(t,-\varepsilon)\right)
\end{aligned}
\tag{6.13}
$$

and

$$
\frac{\partial}{\partial t}\frac{\partial^2 f}{\partial v^2}(t,0) \;=\; \mathfrak{B}_1(t)\frac{\partial^2 f}{\partial v^2}(t,0) + \mathfrak{B}_2(t)\,, \tag{6.14}
$$

where

$$
\mathfrak{B}_1(t) \;=\; \left(2f(t,0)\int\limits_{-\varepsilon}^{\varepsilon} f(t,w)\,\mathrm{d}w - \int\limits_{-\varepsilon}^{\varepsilon} f^2(t,w)\,\mathrm{d}w\right),
$$

and

$$
\begin{aligned}
&\mathfrak{B}_2(t) \;=\\
&= f(t,0)\left(\tfrac{\partial f}{\partial v}(t,\varepsilon)\left(f(t,0) - 2f(t,\varepsilon)\right) - \tfrac{\partial f}{\partial v}(t,-\varepsilon)\left(f(t,0) - 2f(t,-\varepsilon)\right)\right).
\end{aligned}
$$

Therefore,

$$
\begin{aligned}
&\tfrac{\partial^2 f}{\partial v^2}(t,0) = \tfrac{\partial^2 f}{\partial v^2}(0,0)\exp\left(\int\limits_0^t \mathfrak{B}_1(s)\,\mathrm{d}s\right) +\\
&+ \int\limits_0^t \mathfrak{B}_2(s)\exp\left(\int\limits_s^t \mathfrak{B}_1(s')\,\mathrm{d}s'\right)\mathrm{d}s\,.
\end{aligned}
\tag{6.15}
$$

We start by examining the case of Eq. (6.5) and Eq. (6.6). It follows

$$
\frac{\partial f}{\partial v}(t,\varepsilon) \;<\; 0
$$

and

$$
\frac{\partial f}{\partial v}(t,-\varepsilon) \;>\; 0\,,
$$

for $t \in (0,\bar{\tau})$, when $\bar{\tau} \in (0,\tau)$ is sufficiently small.

If Eq. (6.5) is satisfied, then

$$
\mathfrak{B}_2 \;>\; 0 \qquad \text{on} \qquad (0,\bar{\tau})\;.
$$

Additionally,

$$
\mathfrak{B}_1(t) \;=\; \int\limits_{-\varepsilon}^{\varepsilon} f(t,w)\left(2f(t,0) - f(t,w)\right)\mathrm{d}w > 0
$$

on a sufficiently small interval again represented by $(0,\bar{\tau})$.

Next it follows

$$
\frac{\partial^2 f}{\partial v^2}(0,0) \;=\; 0\,,
$$

and

$$\frac{\partial^2 f}{\partial v^2}(t,0) = \int_0^t \mathfrak{B}_2(s) \exp\Big(\int_s^t \mathfrak{B}_1(s')\,\mathrm{d}s' \Big)\,\mathrm{d}s > 0 \qquad t \in (0\,,\,\bar{\tau})\,. \tag{6.16}$$

We can observe by Eq. (6.12) that at $v = 0$, the function $f = f(t, v)$ has its minimum for any $t \in (0, \bar{\tau})$, unlike when for $t = 0$ the maximum is reached.

We see that if

$$0 < \bar{v} = \bar{v}(t) \leq 2\,\varepsilon$$

is such that $f(t,\,.\,)$ attains its maximum at $\bar{v}$, for $t > 0$, then, by continuity, $\bar{v}$ is an increasing function of $t \in (0, \bar{\tau})$. Consequently, this leads to the statement of the first part of Theorem 13.

We examine now the case of Eq. (6.8). If Eq. (6.9) is satisfied, we then obtain

$$\frac{\partial^2 f}{\partial v^2}(t,0) < 0\,, \qquad \forall\, t \in (0, \bar{\tau})\,,$$

on some sufficiently small time interval $(0, \bar{\tau})$, where $\bar{\tau} > 0$, and $v = 0$ continue to be a maximum point. Taking into account Eq. (6.13), we obtain

$$\mathfrak{B}_2(t) < 0\,, \qquad \frac{\partial^2 f}{\partial v^2}(t,0) < 0\,, \qquad \forall\, t \in (0, \tau)\,. \tag{6.17}$$

If Eq. (6.10) is fulfilled by Eq. (6.16) we again observe that $\mathfrak{B}$ is negative. Thus, Eq. (6.17) is satisfied. In both cases, Eq. (6.17) together with Eq. (6.12) are satisfied. This leads to the conclusion that $v = 0$ is a maximum point which completes the proof. ■ QED

By Eq. (6.4) we observe that if Eq. (6.17) holds; hence the value $f(t, 0)$ can only increase for $t \in [0, \tau)$.

Item 1 of Theorem 13 provides sufficient conditions for the creation of a bimodal distribution when the starting state is an unimodal distribution. We can conjecture that the statement, which is limited to a certain short interval of time, can be extended to the entire interval $[0, \tau)$.

Item 2 of Theorem 13 states the conditions for the preservation of a unimodal distribution throughout the interval of existence $[0, \tau)$.

The conditions formulated above lack generality. However, the formulated results indicate interesting and diverse solutions behaviours, which may have both mathematical significance and correspond to the described reality. As we already mentioned, the possibility of blow–ups in a bimodal or unimodal way can be related to the corresponding types of self–organization.

The following example; cf. Ref. [139],

$$\overset{\circ}{f}(v) = \max\{0\,,\,1 - v^4\}\,. \tag{6.18}$$

shows that if

$$\varepsilon < \frac{1}{\sqrt[4]{2}}\,,$$

or

$$\varepsilon > \frac{1}{\sqrt[4]{2}}\,,$$

then we arrive at either a bimodal or an unimodal solution, respectively.

Chapter 7

PDE limits

In this chapter, we consider both $\zeta = 1$ and $\zeta = -1$ in Eq. (2.19) that is both Eq. (2.4) and Eq. (2.18), respectively.

An interesting question is to find mathematical relationships between the solutions of Eq. (2.19) and the solutions of related partial differential equations, which are used most often to describe natural phenomena. This issue can, for example, be formulated as considering a reduction of the range of parameters defining the order of interactions of the system's agents. We can expect that when the appropriate parameters defining the ranges of interactions converge to 0, the solutions of the integro–differential equations converge to the solutions of the appropriate partial differential equations. Such results suggest that Eq. (2.19) can be a direct generalization of models expressed in terms of partial differential equations (PDEs). It may have an essential impact on the modelling process when the models in terms of PDEs are introduced in a rather heuristic way. Their mathematical relationships with the integro–differential equations based on interactions can lead to a justification for them. Moreover, we can expect that in this way we will obtain new possibilities for describing reality in situations where the appropriate parameters defining the interactions are not small enough. An additional value is the lack of need to refer to the differentiability properties of the obtained solutions.

7.1 Symmetric rate of interactions

In what follows in this section, let $\mathbb{V}$ be either $\mathbb{R}^{\mathfrak{n}}$ or $\mathbb{T}^{\mathfrak{n}}$, where the latter is the $\mathfrak{n}$–dimensional torus in $\mathbb{R}^{\mathfrak{n}+1}$. In the case of $\mathbb{T}^{\mathfrak{n}}$, we adhere to the convention that all mathematical operations are understood in the sense of $\mathbb{T}^{\mathfrak{n}}$. We will consider only $\mathbb{V} = \mathbb{R}^{\mathfrak{n}}$, but the arguments can also be applied to $\mathbb{V} = \mathbb{T}^{\mathfrak{n}}$.

The situation for general domains $\mathbb{V}$ is more complicated. In particular, determining the behaviour at the boundary of the domain, in the limit transition, requires additional investigation. This research is crucial to comprehending the dynamics of processes within the life sciences.

The reference [138] dealt with the rate a that is a symmetric function, that is, Eq. (2.7) is satisfied. Equation (2.7) expresses the symmetric nature of interactions between agents: an agent with state v influences another agent with state w in the same way as vice versa.

Let $C_B^m(\mathbb{V})$ be the Banach space of m continuously differentiable functions with the usual norms denoted by $\| \,.\, \|_{[B]}^{(m)}$, $m = 0, 1, 2, \ldots$.

Let $W^{m,p}(\mathbb{V})$ be the classical Sobolev space with the usual norm denoted by $\| \,.\, \|_p^{(m)}$, $m = 0, 1, 2, \ldots$ and $p \geq 1$, cf. Ref. [4].

$X^{(m)} = W^{m,1}(\mathbb{V}) \cap C_B^m(\mathbb{V})$, $m = 0, 1, 2, \ldots$, and $\| \,.\, \|^{(m)}$ is the Banach space equipped with the norm

$$\| \,.\, \|^{(m)} = \| \,.\, \|_1^{(m)} + \| \,.\, \|_{[B]}^{(m)}, \qquad m = 0, 1, 2, \ldots .$$

DOI: 10.1201/9781003537595-7

We simply denote $X = L^1(\mathbb{V}) \cap L^\infty(\mathbb{V})$ equipped with the norm $\|\,.\,\|$.

In case of Eq. (2.18), that is Eq. (2.19) with $\zeta = -1$, the solution $f = f(t)$ is global and the function $t \to \|f(t)\|_p$ decreases for each $p > 1$; see Propositions 1 and 2 in Ref. [138].

Proposition 5 *Let*

- $\gamma > 1$,
- $\overset{\circ}{f}$ *be a probability density such that* $\overset{\circ}{f} \in L^\infty(\mathbb{V})$.

Then, for any $t > 0$, *a solution* $f = f(t)$ *to Eq.* (2.18), *corresponding to the initial datum* $\overset{\circ}{f}$, *exists and is unique in* $L^1 \cap L^\infty$. *The solution* $f = f(t)$ *is a probability density for each* $t > 0$. *Moreover,* $\|f(t)\|_p$ *is bounded for each* $p > 1$ *and each* $t > 0$ *and the function* $t \mapsto \|f(t)\|_p$ *is decreasing. If additionally* $\overset{\circ}{f} \in X^{(m)}$ *and*

$$(a * g) \in X^{(m)} \qquad \text{for each} \quad g \in X^{(m)}\,, \tag{7.1}$$

for some $m = 1, 2, 3, \ldots$, *then the solution* $f = f(t)$ *satisfies* $f(t,\,.\,) \in X^{(m)}$ *for each* $t > 0$.

Proof. The proof of the first part of the statement follows Ref. [138] — Proposition 1. First we observe that the operator on the right-hand side of Eq. (2.18) is locally Lipschitz–continuous in the $L^\infty \cap L^1$-setting. Therefore, the existence and uniqueness of the solution $f = f(t)$, in this setting, at some interval $[0, \tau)$ follows classically. It is obvious that the solution $f = f(t)$ preserves positivity and the L^1–norm of the initial datum for $t \in [0, \tau)$.

By Proposition 2 with $\vartheta(x) = x^p$, $p > 1$, we observe that the solution $f = f(t)$ *a priori* fulfils

$$\frac{\mathrm{d}}{\mathrm{d}t}\int_{\mathbb{V}} f^p(t,v)\,\mathrm{d}v = \frac{\zeta p}{2}\int_{\mathbb{V}}\int_{\mathbb{V}} a(v,w) f(t,v) f(t,w)\,\times$$

$$\times\quad \left(f^{\gamma-1}(t,v) - f^{\gamma-1}(t,w)\right)\left(f^{p-1}(t,v) - f^{p-1}(t,w)\right)\,\mathrm{d}w\,\mathrm{d}v\,,$$

with $\zeta = -1$. The rest of the proof follows the standard application of the Lipschitz continuity in $X^{(m)}$; cf. Proposition 2 in Ref. [138]. Thus, the conclusion follows. ■ QED

In the case of Eq. (2.4), that is Eq. (2.19) with $\zeta = 1$, cf. [138], the solution $f = f(t)$, depending on initial data, can be global ($\tau = \infty$) or local ($\tau < \infty$). It has all bounded L^p–norms on $[0, \tau)$, $p > 1$, for some $\tau > 0$, and the functions $t \mapsto \|f(t)\|_p$ increase for $t \in [0, \tau)$ and $p > 1$. The proof of local existence and the uniqueness result is analogous to the proof of Proposition 5.

7.1.1 Formal diffusive and anti–diffusive limits

In this section, following Ref. [138], we study formally the diffusive limits of the solutions to Eq. (2.19) with both $\zeta = 1$, that is, Eq. (2.4), and $\zeta = -1$, that is, Eq. (2.18). We observe diffusive behaviour for $\zeta = -1$, while for $\zeta = 1$ — anti–diffusive behaviour. Similar effects can be found for example in Ref. [94, 150].

As in Chapter 6, we consider the interaction rate a that has its support "close to the diagonal". We can consider the following simple choice first

$$a(v,w) = \frac{1}{\varepsilon^{\mathrm{n}+2}\omega_{\mathrm{n}}}\chi\left(|v-w| \le \varepsilon\right)\,, \tag{7.2}$$

where

$$\omega_{\mathrm{n}} = \frac{\Gamma(\frac{\mathrm{n}}{2}+1)}{\pi^{\frac{\mathrm{n}}{2}}\,\varepsilon^d}$$

is the normalizing coefficient, that is, $\mathfrak{n}$–dimensional volume of the unit ball centred at 0, $\varepsilon > 0$ is a small parameter and χ is given by Eq. (3.1).

We are interested in the limit when the support of the kernel a tends to the diagonal, but the values grow to infinity. In other words, we are going to study the limit $\varepsilon \to 0$. We note that a given by Eq. (7.2) satisfies Eq. (7.1) and is symmetric; that is, it satisfies Eq. (2.7). The symmetric nature of interactions is essential here. Without it, the result is drastically different — see Section 7.2.

We denote the solution of Eq. (2.19), for fixed $\varepsilon > 0$, by $f^{(\varepsilon)} = f^{(\varepsilon)}(t)$ and assume that it is in $X^{(3)}$. Assuming that $f^{(\varepsilon)}$ tends to f in $X^{(3)}$, as well as $\nabla f^{(\varepsilon)}$ and $D^2 f^{(\varepsilon)}$ tend to ∇f and $D^2 f$, respectively, we see that f is a solution to the following partial differential equation

$$\frac{\partial}{\partial t} f = \zeta\,(1-\gamma)\,\frac{\mathfrak{c}_\mathfrak{n}}{2\,\omega_\mathfrak{n}}\,\nabla\cdot\Big(f^\gamma \nabla f\Big) \tag{7.3}$$

with the positive constant $\mathfrak{c}_\mathfrak{n}$

$$\mathfrak{c}_\mathfrak{n} = \int\limits_{\mathbb{B}(1)} (v_i)^2\,\mathrm{d}v\,, \tag{7.4}$$

where $\mathbb{B}(1)$ is the unit ball, and constant is independent of $i = 1, , \ldots , \mathfrak{n}$. The following standard notation was used: The symbol $\cdot$ denotes the scalar product in $\mathbb{V}$, ∇ is the gradient with respect to v_i, $i = 1, \ldots , \mathfrak{n}$ and $D^2 f$ is the Hessian matrix — it consists of second–order derivatives with respect to v_i, v_j, $i, j = 1, \ldots , \mathfrak{n}$, that is,

$$(\nabla f)_i = \frac{\partial}{\partial v_i} f\,, \qquad \left(D^2 f\right)_{ij} = \frac{\partial^2}{\partial v_i\,\partial v_j} f\,.$$

Taking $\gamma > 1$ as well as $\zeta = -1$, Eq. (7.3) is the classical porous medium equation with a coefficient $(\gamma - 1)$; see, e.g., Refs. [15, 39, 207] and references therein. On the other hand, if $\gamma > 1$ and $\zeta = 1$, it formally results in a nonlinear anti–diffusion equation. A linear anti–diffusion equation was also obtained in a hydrodynamic limit in Ref. [23] — see (4.19) and (4.20) therein.

If solutions are sufficiently regular, i.e., C^2 with respect to v, Eq. (7.3) is equivalent to

$$\frac{\partial}{\partial t} f = \frac{\zeta\,\mathfrak{c}_\mathfrak{n}}{2\,\omega_\mathfrak{n}}\Big(f^\gamma \Delta f - f\,\Delta\left(f^\gamma\right)\Big)\,, \tag{7.5}$$

where Δ is the classical Laplace operator,

$$\Delta f = \frac{\partial}{\partial v}\cdot\frac{\partial}{\partial v} f\,.$$

In fact, we have

$$\begin{aligned} (1-\gamma)\,\nabla\cdot\left(f^\gamma\,\nabla f\right) &= \gamma(1-\gamma) f^{\gamma-1}\nabla f\cdot\nabla f + (1-\gamma)\,f^\gamma \Delta f = \\ = \ & f^\gamma\,\Delta f - f\left(\gamma\,(\gamma-1)\,f^{\gamma-2}\,\nabla f\cdot\nabla f + \gamma f^{\gamma-1}\,\Delta f\right) = \\ = \ & f^\gamma\,\Delta f - f\left(\nabla\cdot\left(\gamma\,f^{\gamma-1}\,\nabla f\right)\right) = f^\gamma \Delta f - f\,\Delta\left(f^\gamma\right)\,. \end{aligned}$$

We can generalize the above argument to Eq. (2.19) with a more general kernel a, see Ref. [138]. We do not need to assume the probability density case, but we can consider generalized moments, as well as the generalized covariance matrix, in a standard way. We need the following assumption; see Ref. [138].

Assumption 7 *We assume*

$$a(u,v) = \frac{1}{\varepsilon^{\mathfrak{n}+2}} \bar{a}\left(\frac{u-v}{\varepsilon}\right), \tag{7.6}$$

where $\varepsilon > 0$ is a small parameter.

1. *The function $\bar{a} : \mathbb{V} \to \mathbb{R}_+$ is bounded, non–negative, and symmetric, that is, $\bar{a}(v) = \bar{a}(-v)$;*

2. *The first, second, third and fourth generalized moments are bounded, that is*

$$\int_{\mathbb{V}} |v|^p \, \bar{a}(v) \, dv < +\infty \quad \text{for} \quad p = 1,2,3,4;$$

3. *The matrix*

$$\Sigma = [\Sigma_{ij}]_{i,j=1,\dots,\mathfrak{n}} = \left[\int_{\mathbb{V}} v_i \, v_j \, \bar{a}(v) \, dv\right]_{i,j=1,\dots,\mathfrak{n}}$$

is positively definite.

We are going to consider the limit $\epsilon \to 0$. The solution of integro–differential equation (2.19) formally *results in*, cf. [138], the following nonlinear partial differential equations (PDEs)

$$\frac{\partial}{\partial t} f(t,v) = \zeta \, \frac{1-\gamma}{2(\gamma+1)} \sum_{i,j=1,\dots,\mathfrak{n}} \Sigma_{ij} \frac{\partial^2 \left(f^{\gamma+1}\right)}{\partial v_i \, \partial v_j}. \tag{7.7}$$

Equation (2.19) for $\zeta = 1$ reduces to Eq. (2.4) whereas for $\zeta = -1$ to Eq. (2.18). In the former case taking $\gamma > 1$ we observe that the resulting partial differential equation (PDE) is of an anti–diffusive type. In this case, achieving any rigorous asymptotic result appears to be quite challenging. On the other hand, Eq. (2.18), for $\gamma > 1$, the outcome is a nonlinear equation for porous media that is widely recognized in the literature; cf. Ref. [15, 39, 207] and references therein, and is supported by a well–established mathematical theory.

If $\bar{a}$ is a probability density, then the generalized moments and the generalized covariance of $\bar{a}$ become the classical moments and the covariance of $\bar{a}$. Assumption 7 holds for a in Eq. (7.2) if $\bar{a}$ is given by

$$\bar{a}(v) = \frac{1}{\omega_d} \chi\left(|v| \le 1\right), \tag{7.8}$$

where χ is given by Eq. (3.1).

Assuming such a form of $\bar{a}$, we use Eq. (7.6) and we get Eq. (7.3) which is formally equivalent to Eq. (7.5).

Our goal is to derive the generalized version of Eq. (7.5); see Ref. [138],

$$\frac{\partial}{\partial t} f(t,v) = \frac{\zeta}{2} \Big\{ f^{\gamma}(t,v) \, \Sigma_{\bar{a}} \cdot D^2 f(t,v) - f(t,v) \, \Sigma_{\bar{a}} \cdot D^2 f^{\gamma}(t,v) \Big\}. \tag{7.9}$$

The symbol $\cdot$ now denotes the scalar product of two $\mathfrak{n} \times \mathfrak{n}$–matrices.

In order to obtain Eq. (7.9) we apply the Taylor expansions

$$\begin{aligned} f^{(\varepsilon)}(t,w) &= f^{(\varepsilon)}(t,v) + (w-v) \cdot \nabla f^{(\varepsilon)}(t,v) + \\ &+ \tfrac{1}{2}(w-v) \cdot D^2 f^{(\varepsilon)}(t,v)(w-v) + \mathcal{O}\left(|w-v|^3\right), \end{aligned}$$

and

$$\left(f^{(\varepsilon)}\right)^\gamma(t,w) = \left(f^{(\varepsilon)}\right)^\gamma(t,v) + (w-v)\cdot\nabla\left(f^{(\varepsilon)}\right)^\gamma(t,v) + \\ + \tfrac{1}{2}(w-v)\cdot D^2\left(f^{(\varepsilon)}\right)^\gamma(t,v)(w-v) + \mathcal{O}\left(|w-v|^3\right).$$

Inserting these expressions into the right–hand side of Eq. (2.19) and considering Eq. (7.6) results in

$$\begin{aligned}
&\zeta\int_{\mathbb{V}}\Big(a(v,w)f^{(\varepsilon)}(t,w)\left(f^{(\varepsilon)}\right)^\gamma(t,v)\\
-&\; a(v,w)f^{(\varepsilon)}(t,v)\left(f^{(\varepsilon)}\right)^\gamma(t,w)\Big)\,\mathrm{d}w =\\
=&\;\tfrac{\zeta}{\varepsilon^{n+2}}\int_{\mathbb{V}}\bar{a}\left(\tfrac{v-w}{\varepsilon}\right)\left(f^{(\varepsilon)}\right)^\gamma(t,v)\Big(f^{(\varepsilon)}(t,v)+\\
+&\;\nabla f^{(\varepsilon)}(t,v)\cdot(w-v)+\\
+&\;\tfrac{1}{2}(w-v)\cdot D^2 f^{(\varepsilon)}(t,v)(w-v)+\mathcal{O}\left(|w-v|^3\right)\Big)\,\mathrm{d}w\\
-&\;\tfrac{\zeta}{\varepsilon^{n+2}}\int_{\mathbb{V}}\bar{a}\left(\tfrac{v-w}{\varepsilon}\right)f^{(\varepsilon)}(t,v)\Big(\left(f^{(\varepsilon)}\right)^\gamma(t,v)+\\
+&\;\nabla\left(f^{(\varepsilon)}\right)^\gamma(t,v)\cdot(w-v)+\\
+&\;\tfrac{1}{2}(w-v)\cdot D^2\left(f^{(\varepsilon)}\right)^\gamma(t,v)(w-v)+\mathcal{O}(|w-v|^3)\Big)\,\mathrm{d}w\,.
\end{aligned} \tag{7.10}$$

Due to the symmetry of $\bar{a}$, the first-order terms in $w-v$ cancel out. Then, the second–order terms lead to

$$\frac{\zeta}{2}\left(\left(f^{(\varepsilon)}\right)^\gamma(t,v)\,\Sigma_{\bar{a}}\cdot D^2\left(f^{(\varepsilon)}\right)(t,v) - f^{(\varepsilon)}(t,v)\,\Sigma_{\bar{a}}\cdot D^2\left(f^{(\varepsilon)}\right)^\gamma(t,v)\right).$$

The limit $\varepsilon\to 0$ formally results in the diffusion or anti–diffusion equation (7.9).

In the particular case of Eq. (7.2), all mixed generalized second moments of $\bar{a}$ vanish. Then, by Eq. (7.4), the pure second generalized moments are detailed below

$$\begin{aligned}
&\zeta\int_{\mathbb{V}}\Big(a(w,v)f^{(\varepsilon)}(t,w)\left(f^{(\varepsilon)}\right)^\gamma(t,v)\\
-&\; a(w,v)f^{(\varepsilon)}(t,v)\left(f^{(\varepsilon)}\right)^\gamma(t,w)\Big)\,\mathrm{d}w =\\
=&\;\tfrac{\zeta\, c_n}{2\,\omega_n}\Big(\left(f^{(\varepsilon)}\right)^\gamma(t,v)\nabla^2 f^{(\varepsilon)}(t,v)\\
-&\; f^{(\varepsilon)}(t,v)\,\nabla^2\left(\left(f^{(\varepsilon)}\right)^\gamma(t,v)\right)\Big) + \mathcal{O}(\varepsilon)\,.
\end{aligned} \tag{7.11}$$

Under that assumption of this section, in the limit $\varepsilon\to 0$, Eq. (2.19) formally results in the diffusion or anti–diffusion equation Eq. (7.5) depending on ζ. Equation (7.5) becomes the classical porous–media equation when $\zeta = -1$. Its theory of existence, uniqueness, and regularity can be found in Ref. [207], see also references therein. No known theory exists for anti–diffusion equations, particularly for Eq. (7.5) with $\zeta = 1$.

7.1.2 Convergence result

In this section, our aim is to prove the rigorous convergence result, namely a diffusive limit for $\zeta = -1$. The result is in some sense in the spirit of the diffusive limit for a class of bilinear equations studied in Ref. [146].

We assume a stronger regularity of the initial data $\overset{\circ}{f}$ and obtain the convergence of order ε^2. The result is not optimal from the point of view of regularity, and of course, if necessary, can be generalized to weaker assumptions.

The symmetry of a allows an increase in the convergence rate from ε^1 to ε^2.

As we already mentioned, a significant difference is observed between cases $\zeta = -1$ (diffusion) and $\zeta = 1$ (anti–diffusion). Namely, solutions to Eq. (7.5) with $\zeta = -1$ exist in some interval $[0, \tau]$ provided that the initial function $\overset{\circ}{f}$ is sufficiently regular. In fact, for Eq. (2.19) with $\zeta = -1$ the global existence and uniqueness theory is given by Proposition 5, and for its diffusive limit (7.3) it follows from the classical theory of porous media; see Ref. [207]. In the case $\zeta = 1$, it is unknown.

We can convert Eq. (7.9) into the porous–media equation and formulate an existence theorem in the diffusive scenario when $\zeta = -1$. Taking into account that the matrix $\Sigma_{\bar{a}}$ is symmetric and positively definite, we see that there exists a $\mathfrak{n} \times \mathfrak{n}$–matrix B such that $\Sigma_{\bar{a}} = B^2$. We can now introduce the function $F(t, v) = f(t, Bv)$ and use the relations between the derivatives

$$
\begin{aligned}
\nabla F(t,v) &= (B\,\nabla f)\,(t, Bv)\;, \\
\Delta F(t,v) &= \left(B^2 \cdot D^2 f\right)(t, Bv) = \left(\Sigma_{\bar{a}} \cdot D^2 f\right)(t, Bv)\;.
\end{aligned}
\tag{7.12}
$$

Equation (7.9) is converted to

$$
\frac{\partial}{\partial t} F = -\frac{1}{2} F^{\gamma}\,\Delta F + \frac{1}{2} F\,\Delta F^{\gamma}\,. \tag{7.13}
$$

The initial data are given by $\overset{\circ}{F}(v) = \overset{\circ}{f}(Bv)$. We have

$$
\begin{aligned}
\nabla \cdot \left(F^{\gamma} \nabla F\right)(t,v) &= \nabla \cdot \left(F^{\gamma}(t,v)\,(B \nabla f)\,(t, Bv)\right) = \\
&= \tfrac{1}{\gamma+1}\, B^2 \cdot D^2 \left(f^{\gamma+1}\right)(t, Bv)\;.
\end{aligned}
\tag{7.14}
$$

Thus we can obtain an equation which is formally equivalent to Eq. (7.9) and reduces to a porous–media equation

$$
\frac{\partial}{\partial t} f = \frac{\gamma - 1}{2(\gamma+1)}\, \Sigma_{\bar{a}} \cdot D^2 \left(f^{\gamma+1}\right)\,. \tag{7.15}
$$

Then $F(t, v) = f(t, Bv)$ satisfies the porous–media equation

$$
\frac{\partial}{\partial t} F = \frac{\gamma - 1}{2(\gamma+1)}\, \Delta \left(F^{\gamma+1}\right)\,. \tag{7.16}
$$

We can now state the following existence result; see Proposition 3 in Ref. [138].

Proposition 6 *Let*

- $\gamma > 1$, $\zeta = -1$,
- $\overset{\circ}{f}$ *be a probability density on* $\mathbb{V}$ *such that* $\overset{\circ}{f} \in X^{(m)}$, $m \geq 1$,
- $\bar{a} \in X$ *satisfy Assumption 7.*

Then Eq. (7.15)*, has a unique solution* $f = f(t)$ *in* $X^{(m)}$*, corresponding to the initial datum* $\overset{\circ}{f}$*, on some interval* $[0, \tau]$, $\tau > 0$.

Proof. Following Ref. [138] we use the substitution $F(t, v) = f(t, Bv)$ to see that Eq. (7.15) becomes Eq. (7.16) and therefore we obtain the Cauchy problem

$$\frac{\partial}{\partial t} F = \frac{\gamma - 1}{2} \nabla \cdot (F^{\gamma} \nabla F) ,$$

with the initial data $\overset{\circ}{F}(u) = \overset{\circ}{f}(Bu)$. We use the known porous–media existence theory to end the proof. ■ QED

The following theorem states the asymptotic result; see Theorem 1 in Ref. [138].

Theorem 14 *Let*

- $\gamma > 1$, $\zeta = -1$,
- $\overset{\circ}{f}$ *be a probability density on* $\mathbb{V}$ *such that* $\overset{\circ}{f} \in X^{(4)}$*, and, moreover, a satisfied Assumption 7 with* $\bar{a} \in X$.

Additionally, let

- $f^{(\varepsilon)} = f^{(\varepsilon)}(t)$ *defined on* $[0, \tau] \times \mathbb{V}$ *be a solution of Eq.* (2.19) *with the initial datum* $\overset{\circ}{f}$;
- $f = f(t)$ *in* X *be a solution of Eq.* (7.15) *with the same initial datum* $\overset{\circ}{f}$.

Then

$$\sup_{[0,\tau]} \|f^{(\varepsilon)} - f\| \leq \mathfrak{c}\, \varepsilon^2 ,$$

with some constant $\mathfrak{c} > 0$.

Proof. See Ref. [138]. Assume that $f^{(\varepsilon)} \in X^{(4)}$ is the solution of Eq. (2.19) with a defined by Eq. (7.6) and corresponding to the initial data $\overset{\circ}{f}$. The successive steps of the proof are following.

1. The starting kinetic equation is changed into an equation that is formally equivalent to the porous–media equation with a small perturbation;
2. The perturbed porous–media equation is transformed to the standard form by the change of variables $v \to Bv$;

3. The comparison principle is applied to the porous–media equation.

Let $(t, v) \in [0, \tau] \times \mathbb{V}$ be fixed. The Taylor expansions applied to Eq. (2.19), as in Section 7.1.1, yields the right–hand side given by Eq. (7.10). Using the symmetry of $\bar{a}$, we observe that the terms of the first and third order of $w - u$ disappear. The second-order terms give

$$I_\varepsilon(f^{(\varepsilon)})(t,v) = \tfrac{-1}{2}\Big\{ \left(f^{(\varepsilon)}\right)^\gamma (t,v)\, \Sigma_{\bar{a}} \cdot D^2 \left(f^{(\varepsilon)}\right)(t,v)$$

$$- \quad f^{(\varepsilon)}(t,v)\, \Sigma_{\bar{a}} \cdot D^2 \left(f^{(\varepsilon)}\right)^\gamma (t,v)\Big\} =$$

$$= \tfrac{\gamma-1}{2(\gamma+1)}\, \Sigma_{\bar{a}} \cdot (D^2 f^{\gamma+1}).$$

Therefore, Eq. (7.10) has the form

$$\int_{\mathbb{V}} \tfrac{1}{\varepsilon^{n+2}} \bar{a}\left(\tfrac{v-w}{\varepsilon}\right) \Big(f^{(\varepsilon)}(t,w) \left(f^{(\varepsilon)}\right)^\gamma (t,v)$$

$$- \quad f^{(\varepsilon)}(t,v) \left(f^{(\varepsilon)}\right)^\gamma (t,w)\Big)\, dw = I_\varepsilon(f^{(\varepsilon)})(t,v) + R_\varepsilon(f^{(\varepsilon)})(t,v),$$

with

$$R_\varepsilon(f^{(\varepsilon)})(t,v) =$$

$$= \int_{\mathbb{V}} \tfrac{1}{\varepsilon^{n+2}} \bar{a}\left(\tfrac{v-w}{\varepsilon}\right) \left(f^{(\varepsilon)}\right)^\gamma (t,v) \times$$

$$\times \int_0^1 \tfrac{(1-\theta)^3}{6} D^4 f^{(\varepsilon)}(t, v+\theta(w-v))\,(w-v)^4\, d\theta\, dw +$$

$$+ \int_{\mathbb{V}} \tfrac{1}{\varepsilon^{n+2}} \bar{a}\left(\tfrac{v-w}{\varepsilon}\right) f^{(\varepsilon)}(t,v) \times$$

$$\times \int_0^1 \tfrac{(1-\theta)^3}{6} D^4 \left(f^{(\varepsilon)}\right)^\gamma (t, v+\theta(w-v))\,(w-v)^4\, d\theta\, dw,$$

and $D^4 \left(f^{(\varepsilon)}\right)^p$ contains all fourth–order partial derivatives

$$\frac{\partial^4}{\partial v_i\, \partial v_j\, \partial v_k\, \partial v_l} \left(f^{(\varepsilon)}\right)^p,$$

and

$$D^4 \left(f^{(\varepsilon)}\right)^p (t, v+\theta(w-v))\,(w-v)^4,$$

is the 4–linear form

$$D^4 \left(f^{(\varepsilon)}\right)^p (t, v+\theta(w-v)),$$

acting on four copies of the vector $w - v$.

Then the remainder $R_\varepsilon(f^{(\varepsilon)})$ can be estimated by

$$\int_{\mathbb{V}} \tfrac{1}{\varepsilon^{n+2}} \bar{a}\left(\tfrac{v-w}{\varepsilon}\right) f^{(\varepsilon)}(t,v) |w-v|^4 \times$$

$$\times \int_0^1 \tfrac{(1-\theta)^3}{6} \left(\|f^{(\varepsilon)}\|_{L^\infty}^{\gamma-1} \|D^4 f^{(\varepsilon)}\|_{L^\infty} + \|D^4 \left(f^{(\varepsilon)}\right)^\gamma\|_{L^\infty}\right) d\theta\, dw,$$

and the X–norm can be estimated by

$$\mathfrak{c}\,\varepsilon^2\,,$$

where $\mathfrak{c}$ is a constant that depends on the fourth moments of $\bar{a}$ and the L^∞–norm of $D^4 f^{(\varepsilon)}$.

Now we aim to estimate the distance between $f^{(\varepsilon)}$ and f. The previous arguments lead to the perturbed porous–media equation for $f^{(\varepsilon)}$

$$\frac{\partial}{\partial t} f^{(\varepsilon)}(t,v) \;=\; \frac{\gamma-1}{2(\gamma+1)}\,\Sigma_{\bar{a}}\cdot D^2\left(f^{(\varepsilon)}\right)^{\gamma+1}(t,x) \;+\; R_\varepsilon(f^{(\varepsilon)})(t,v)\,.$$

In the same way as the transformation $f \to F$, one can define $F^{(\varepsilon)}(t,v) := f^{(\varepsilon)}\,(t,Bv)$, that satisfies the following equation

$$\frac{\partial}{\partial t} F^{(\varepsilon)}(t,v) \;=\; \frac{\gamma-1}{2(\gamma+1)}\,\Delta\left(f^{(\varepsilon)}\right)^{\gamma+1}(t,x) \;+\; R_\varepsilon(f^{(\varepsilon)})(t,Bv)$$

with the initial data

$$F^{(\varepsilon)}(0,v) = \overset{\circ}{F}(v) = \overset{\circ}{f}(Bv)\,.$$

Because F satisfies the porous–media equation — Eq. (7.16) — with the same initial datum, one can apply the classical comparison theorems for porous media; see Refs. [26, 39]. Therefore, it follows

$$\sup_{[0,\tau]} \|F^{(\varepsilon)} - F\| \;\leq\; \mathfrak{c}\,\varepsilon^2\,.$$

Using the fact that the matrix B is not singular, a similar estimate for $f^{(\varepsilon)} - f$ is obtained, which ends the proof. ■ QED

If the mixed generalized moments of $\bar{a}$ vanish, then the covariance matrix $\Sigma_{\bar{a}}$ is diagonal. Then we have the diffusive or anti–diffusive limiting equation, that is, Eq. (7.5), which is a particular case of Eq. (7.9). For $\zeta = -1$, we state this fact in the following corollary; see Corollary 1 in Ref. [138].

Corollary 5 *Let*

- $\gamma > 1$, $\zeta = -1$,
- $\overset{\circ}{f}$ *be a probability density on* $\mathbb{V}$ *such that* $\overset{\circ}{f} \in X^{(4)}$,
- a *given by Eq.* (7.2)

Additionally, let

- $f^{(\varepsilon)} = f^{(\varepsilon)}(t,v)$, *defined on* $[0,\tau]\times\mathbb{V}$, *be a solution of Eq.* (2.19), *corresponding to the initial datum* $\overset{\circ}{f}$,
- $f = f(t)$, *defined for* $t \in [0,\tau]$, *be a solution in* X *of Eq.* (7.3), *corresponding to the same initial data* $\overset{\circ}{f}$.

Then

$$\sup_{[0,\tau]} \|f^{(\varepsilon)} - f\| \;\leq\; \mathfrak{c}\,\varepsilon^2\,,$$

with some constant $\mathfrak{c} > 0$.

Proof. We follow Ref. [138]. Assume Eq. (7.2). The covariance matrix $\Sigma_{\bar{a}}$ is proportional to the unit matrix I and the matrix operation

$$\Sigma_{\bar{a}} \cdot D^2 f$$

gives

$$\frac{c_n}{\omega_n} \Delta f \,.$$

Then we conclude by observing that Eq. (7.3) is a particular case of Eq. (7.15). ■ QED

The results presented in this section show that the general equation Eq. (2.1), in particular cases, can describe the processes of an organization or the tendency to order — Eq. (2.19) or disorganization — Eq. (2.18). In general situations, without imposing specific conditions, both scenarios may occur. This is a topic for further mathematical investigations with a significant potential for enhancing the characterization of particular processes in the life and social sciences.

The concept of organization in this context can refer to focusing on concentration on specific states.

Additionally, it is observed that the fundamental mathematical theory, such as the properties of existence, regularity, or uniqueness, for integro–differential kinetic equations may be more satisfactory than that for the associated partial differential equations.

7.2 Asymmetric rate of interactions

In the previous section, we considered the symmetric interactions between the agents of a system. However, a variety of processes in the life and social sciences can be characterized by asymmetric relationships and therefore asymmetric interactions. The reciprocal action between a psychologist or psychiatrist and a patient, a teacher, and a student are examples of asymmetric interactions. Very important situations regarding vaccination issues that take into account the psychological individual aspects can be further examples; see Refs. [25, 79, 80, 154, 210, 213]. Cell divisions can lead to two daughter cells with different fates; see Ref. [202]. In economics, an intriguing illustration of asymmetric information on a market is the lemon–cherries theory; see Ref. [8], in which one agent has more knowledge than the others. Usually, it can lead to market degradation. It should be emphasized that the situations given above are still open to modelling in the language of interactions of agents considered in this monograph.

Sometimes, however, observed asymmetric behaviour can be directly connected with an asymmetric manner of agent interactions. This is a common occurrence in many scenarios in ecology in which agent interactions are of asymmetrical nature; see, e.g., Refs. [2, 20]. Similarly, in the case of opinion dynamics, we should expect that extreme opinions can have a different impact on moderate opinions than vice versa. This applies, in particular, to political opinions. These aspects also require further research.

In this section, we refer to asymmetric behaviour through asymmetric rate of interactions with the interpretation that the state of one agent influences the state of another differently than the other way around. Some examples of possible applications are given in Chapter 11, Sections 12.3 and 13.2.3.

In the asymmetric case the formal analysis, see the next subsection, leads to the nonlinear hyperbolic equation, namely the Burgers–type equation, inviscid Burgers equation; cf. Section 7.2.3.

As previously; cf. Section 7.1, let $\mathbb{V}$ be either $\mathbb{R}^{\mathfrak{n}}$ or $\mathbb{T}^{\mathfrak{n}}$, where the latter is the $\mathfrak{n}$–dimensional torus in $\mathbb{R}^{\mathfrak{n}+1}$. We will consider only $\mathbb{V} = \mathbb{R}^{\mathfrak{n}}$, but the arguments can also be applied to $\mathbb{V} = \mathbb{T}^{\mathfrak{n}}$.

7.2.1 Formal limit

Following Ref. [134] we investigate Eq. (2.4) with $a(w, v) = \tilde{a}(w - v)$ and, for simplicity, we omit the tilde. We introduce a small parameter $\varepsilon > 0$ and consider the following equation

$$\begin{aligned} \frac{\partial}{\partial t} f(t,v) &= \frac{1}{\varepsilon^{\mathfrak{n}+\nu}} \left(f^{\gamma}(t,v) \int_{\mathbb{V}} a\left(\tfrac{w-v}{\varepsilon}\right) f(t,w)\,\mathrm{d}w \right. \\ &- \left. f(t,v) \int_{\mathbb{V}} a\left(\tfrac{v-w}{\varepsilon}\right) f^{\gamma}(t,w)\,\mathrm{d}w \right), \end{aligned} \tag{7.17}$$

where the parameter ε effectively captures the "shrinking" impact on the range of interactions, similarly to Chapter 6, and ν takes the value either 0 or 1.

We introduce the new variable as follows

$$w' = \frac{w - v}{\varepsilon}$$

and note that the expression on the right–hand side of Eq. (7.17), denoted by $\mathcal{R}_{\varepsilon}(f)$, can be expressed as

$$\begin{aligned} \mathcal{R}_{\varepsilon}(f)(t,v) = \frac{1}{\varepsilon^{\nu}} & \left(f^{\gamma}(t,v) \int_{\mathbb{V}} a(w')\, f(t, v + \varepsilon w')\,\mathrm{d}w' \right. \\ - \; & \left. f(t,v) \int_{\mathbb{V}} a(-w')\, f^{\gamma}(t, v + \varepsilon w')\,\mathrm{d}w' \right). \end{aligned} \tag{7.18}$$

Thus, Eq. (7.17) reads

$$\frac{\partial}{\partial t} f(t,v) = \mathcal{R}_{\varepsilon}(f)(t,v)\,. \tag{7.19}$$

In the remainder of this chapter, we adopt the following assumption.

Assumption 8 *The function a is nonnegative integrable and such that*

$$\begin{aligned} \int_{\mathbb{V}} a(w)\,\mathrm{d}w &= \int_{\mathbb{V}} a(-w)\,\mathrm{d}w\,, \\ \int_{\mathbb{V}} a(w)\, w_i\,\mathrm{d}w &= -\int_{\mathbb{V}} a(-w)\, w_i\,\mathrm{d}w\,, \qquad i = 1, \ldots, \mathfrak{n}, \end{aligned} \tag{7.20}$$

are bounded. We denote

$$\begin{aligned} B_0 &= \int_{\mathbb{V}} a(w)\,\mathrm{d}w\,, \\ B_{1,i} &= \int_{\mathbb{V}} a(w)\, w_i\,\mathrm{d}w\,, \qquad i = 1, \ldots, \mathfrak{n}. \end{aligned} \tag{7.21}$$

A typical example of such a kernel a in the 1–dimension ($\mathfrak{n}=1$) is the normalized indicator function of the interval $[0,1]$.

It follows that

$$\mathcal{R}_\varepsilon(f)(t,v) = \frac{1}{\varepsilon^\nu}\Bigg(f^\gamma(t,v)\int_{\mathbb{V}} a\,(w')\,(f\,(t,v+\varepsilon w') - f(t,v))\,dw'$$

$$- \quad f(t,v)\int_{\mathbb{V}} a\,(-w')\,(f^\gamma\,(t,v+\varepsilon w') - f^\gamma(t,v))\,dw' \Bigg)$$

and asymptotically, the behaviour of $\mathcal{R}_\varepsilon(f)$ is like $\left(\frac{1}{\varepsilon}\right)^{\nu-1}\,\mathcal{R}(f)(t,v)$, where

$$\mathcal{R}(f)(t,v) = f^\gamma(t,v)\int_{\mathbb{V}} a\,(w')\,\frac{\partial}{\partial v}f(t,v)\cdot w'\,dw'$$

$$- \quad f(t,v)\int_{\mathbb{V}} a\,(-w')\,\frac{\partial}{\partial v}\,f^\gamma(t,v)\,\cdot w'\,dw'\,,$$

where, as before, $\cdot$ denotes the scalar product in $\mathbb{R}^\mathfrak{n}$.

In the limit, the following Generalized Inviscid Burgers equation is formally obtained

$$\frac{\partial}{\partial t}f(t,v) = \varepsilon^{1-\nu}\,(\gamma+1)\,\left(f(t,v)\right)^\gamma\,B_1\cdot\frac{\partial}{\partial v}f(t,v)\,. \tag{7.22}$$

in which for $\nu=0$ the parameter ε appears explicitly. Alternatively, if $\nu=1$ in Eq. (7.17), the corresponding version of Eq. (7.22) does not include ε. This is somehow similar to what is observed in the hydrodynamic limits of kinetic equations; cf. [23, 24] and references therein.

Expanding $\mathcal{R}_\varepsilon(f)(t,v)$ yields

$$\begin{aligned}
\mathcal{R}_\varepsilon(f)(t,v) &= \\
&= \varepsilon^{1-\nu}\sum_{i=1}^{\mathfrak{n}} B_{1i}\left(f^\gamma(t,v)\frac{\partial f(t,v)}{\partial v_i} + f(t,v)\frac{\partial f^\gamma(t,v)}{\partial v_i}\right) + \\
&+ \frac{\varepsilon^{2-\nu}}{2!}\sum_{i,j=1,\dots,\mathfrak{n}} B_{2ij}\left(f^\gamma(t,v)\,\frac{\partial^2 f(t,v)}{\partial v_i\,\partial v_j} - f(t,v)\frac{\partial^2 f^\gamma(t,v)}{\partial v_i\,\partial v_j}\right) + \\
&+ \frac{\varepsilon^{3-\nu}}{3!}\sum_{i,j,k=1,\dots,\mathfrak{n}} B_{3ijk}\Bigg(f^\gamma(t,v)\,\frac{\partial^3 f(t,v)}{\partial v_i\,\partial v_j\,\partial v_k} + \\
&+ f(t,v)\frac{\partial^3 f^\gamma(t,v)}{\partial v_i\,\partial v_j\,\partial v_k}\Bigg) + \dots,
\end{aligned} \tag{7.23}$$

where

$$B_{2ij} = \int_{\mathbb{V}} w'_i w'_j\,a(w')\,dw' \qquad \text{and} \quad B_{3ijk} = \int_{\mathbb{V}} w'_i w'_j w'_k\,a(w')\,dw'\,,$$

$i,j,k = 1\,\dots\,\mathfrak{n}$.

It is evident that when $\gamma=1$ the second–order term disappears just as any other even–order term. Conversely, when $\gamma>1$, the second-order term results in an anti-diffusive scenario. Hence, the situation of $\gamma>1$ is inherently singular. As we shall see; cf. Remark 3, the case $\gamma=1$ can also have a singular nature.

7.2.2 Asymmetric interactions

Considering Eq. (7.17) with $\gamma = 1$ and taking a which is not a symmetric function of its variables, we observe that Eq. (7.17) is no more trivial, in contrast to the symmetric case. In case $\gamma = 1$, we can apply the natural approach in the $L^1 = L^1(\mathbb{V})$ space. As we have already discussed, this setting does not make direct mathematical sense for $\gamma > 1$.

The operator $\mathcal{R}_\varepsilon$ is Lipschitz–continuous in L^1, that is,

$$\|\mathcal{R}_\varepsilon(f) - \mathcal{R}_\varepsilon(g)\|_1 \leq \mathfrak{c}\Big(\|f\|_1 + \|g\|_1\Big)\|f - g\|_1\,, \quad \forall\, f, g \in L^1\,, \tag{7.24}$$

where, as before, $\|\,.\,\|_1$ is the norm in L^1 and $\mathfrak{c}$ is a positive constant. Furthermore, the operator is conservative in the sense

$$\int_{\mathbb{V}} \mathcal{R}_\varepsilon(f)(v)\, \mathrm{d}v = 0\,, \quad \forall\, f \in L^1\,. \tag{7.25}$$

Let $a \in L^\infty(\mathbb{V})$ be nonnegative. The local (in time) result on the existence and uniqueness of solutions in L^1 then follows from Eq. (7.24). It is easy to see, by the form of Eq. (7.19), that the solution $f = f(t)$, for each $t > 0$, is *a priori* nonnegative and according to Eq. (7.25), its L^1–norm is conserved for the nonnegative initial datum $f(0) = \overset{\circ}{f}$ in L^1. It follows that the solution can be extended to any $t > 0$. Therefore, the global–in–time existence result holds; cf. Ref. [134]. We have; see Ref. [134],

Proposition 7 *Let*

- $\gamma = 1$, $\varepsilon > 0$,
- $\nu = 0$ *or* 1,
- $\overset{\circ}{f}$ *be a probability density,*
- $a \in L^\infty$.

Then, for any $t > 0$, *there exists a unique solution* $f = f(t)$ *of Eq. (7.19) in* L^1, *corresponding to the initial data* $\overset{\circ}{f}$. *The function* $f = f(t)$ *is a probability density for any* $t > 0$.

Now, we examine $\gamma \geq 1$ mainly interested in the case $\gamma > 1$. We are going to prove the existence of $\tau > 0$, which is independent of ε, such that there exists a unique solution to Eq. (7.17) on $[0, \tau]$. As previously, the operator $\mathcal{R}_\varepsilon$ is Lipschitz–continuous in $L^\infty(\mathbb{V})$, so the local existence–uniqueness result holds. It is evident that when $\nu = 0$, the result of existence and uniqueness can be established in $[0, \tau]$, where τ is independent of ε; see Proposition 2 in Ref. [134].

Proposition 8 *Let*

- $\gamma \geq 1$, $\varepsilon > 0$,
- $\nu = 0$ *or* 1,

- $a \in L^1(\mathbb{V})$,
- $\overset{\circ}{f}$ *be a probability density such that* $\overset{\circ}{f} \in L^\infty(\mathbb{V})$.

Then there exists $\tau > 0$, *which is independent of* ε, *such that, in the time interval* $[0\,,\,\varepsilon^\nu\,\tau]$, *there exists a unique solution* $f = f(t)$ *of Eq. (7.19) in* $L^1 \cap L^\infty$, *corresponding to the initial data* $\overset{\circ}{f}$. *The function* $f = f(t)$ *is a probability density for any* $t \in [0\,,\,\varepsilon^\nu\,\tau]$.

Proof. We follow Ref. [134]. The operator $\mathcal{R}_\varepsilon$, in Eq. (7.19), satisfies the (local) Lipschitz condition in both L^∞ and $L^1 \cap L^\infty$. We are going to find an estimate *a priori* of the solution. For this purpose, we establish the integral inequality

$$\|f(t)\|_\infty \leq \left\| \overset{\circ}{f} \right\|_\infty + \frac{2}{\varepsilon^\nu} B_0 \int_0^t \|f(s)\|_\infty^{\gamma+1}\, ds\,.$$

By the comparison theorem we have

$$\|f(t)\|_\infty \leq \frac{\left\| \overset{\circ}{f} \right\|_\infty}{(1 - \mathfrak{c}_0 \varepsilon^{-\nu} t)^{\frac{1}{\gamma}}}$$

with some constant $\mathfrak{c}_0$ depending on $\| \overset{\circ}{f} \|_\infty$. This concludes the proof. ■ QED

The Lipschitz constant for $\nu = 1$ depends on ε, and the time interval of the existence of solution can also depend. However, if we show a ε–independent *a priori* estimate for a solution, we obtain the existence and uniqueness in a time interval independent of ε.

Returning to $\nu = 0$, we write equations for the derivatives of the solution given by (8). Keeping in mind Eq. (7.18) with $\nu = 0$, we can examine the smoothness of the solutions.

We use the notation already introduced in Section 7.1, that is, $X^{(m)} = W^{m,1}(\mathbb{V}) \cap C_B^m(\mathbb{V})$, $m = 0, 1, 2, \ldots$, and $X = X^{(0)} = L^1 \cap L^\infty$.

Following Ref. [134], we have

Proposition 9 *Let*

- $\gamma \geq 1,\ \varepsilon > 0$,
- $\nu = 0$ *or* 1,
- $a \in L^\infty$;
- $\overset{\circ}{f}$ *be a probability density such that* $\overset{\circ}{f} \in X^{(m)}$ *and* $a \in X^{(m)}$ *for some* $m = 1, 2, 3, \ldots$.

Then there exists $\tau > 0$ *such that the solution* $f = f(t)$, *corresponding to the initial data* $\overset{\circ}{f}$ *and given by Proposition 8, satisfies* $f(t\,,\,.) \in X^{(m)}$ *for all* $t \in [0\,,\,\varepsilon^\nu\,\tau]$. *If* $\nu = 0$ *then*

$$\|f(t\,,\,.)\|^{(m)} \leq \mathfrak{c} \qquad \forall\, t \in [0, \tau]\,,$$

where $\mathfrak{c}$ *is a positive constant.*

Proof. The application of Lipschitz–continuity in the $X^{(m)}$ spaces yields the statement. ■ QED

7.2.3 Burgers equation

We fix our attention on the following Partial Differential Equation that represents a version of the Generalized Inviscid Burgers equation

$$\frac{\partial}{\partial t} f(t,v) = \varepsilon^{1-\nu}\, \sigma(v) \cdot \left(f^{\gamma}(t,v)\, \frac{\partial}{\partial v} f(t,v) \right) . \tag{7.26}$$

where $\sigma = \sigma(v)$ is a function $\mathbb{V} \to \mathbb{V}$.

Setting $\nu = 1$, $\sigma \equiv 1$ and $\gamma = 1$, one gets the classical Inviscid Burgers equation. In the general case, Eq. (7.26) does not have to be conservative, in contrast to Eq. (7.22). However, the constant function σ leads to the conservative equation.

We have the following theorem; see Theorem 1 in Ref. [134].

Theorem 15 *Let σ and $\overset{\circ}{f}$ be non–negative, bounded C^1 functions on $\mathbb{V}$, whose derivatives are bounded.*

1. *There exist $\tau > 0$ and a unique solution f to Eq. (7.26) on the interval $[0, \frac{\tau}{\varepsilon^{1-\nu}}]$. The function f is C^1, bounded, and nonnegative.*
2. *If additionally $\frac{\partial^2}{\partial v^2} \overset{\circ}{f}{}^{p}$, for some $p \geq 1$ and $\frac{\partial^2}{\partial v^2}\sigma$ are bounded and continuous, then f^p is C^2 with bounded derivatives $\frac{\partial^2}{\partial v^2} f^p(t,v)$ on the interval $[0, \frac{\tau}{\varepsilon^{1-\nu}}]$.*
3. *If $\overset{\circ}{f} \in L^1$, then the solution $f(t,.)$ is in L^1 on the interval $[0, \frac{\tau}{\varepsilon^{1-\nu}}]$.*

Proof. We follow Ref. [134]. Assume that $\nu = 1$. To simplify notation, we consider the one–dimensional scenario $\mathfrak{n} = 1$, but the proof follows analogously also for any $\mathfrak{n} > 1$.

Given $\overset{\circ}{f}$. Characteristics $\eta(t) = \eta(t,v)$ are solutions to the Cauchy problem for the ordinary differential equation with the variable $v \in \mathbb{V}$ treated as a parameter

$$\begin{aligned} \frac{d\eta(t)}{dt} &= -\sigma(\eta(t))\, \overset{\circ}{f}{}^{\gamma}(v) , \\ \eta(0) &= v . \end{aligned}$$

Because $\overset{\circ}{f}$ is bounded and σ is Lipschitz–continuous, the solutions, that is, the characteristics $\eta = \eta(t,v)$ exist globally. The implicit formula expresses the solution to Eq. (7.26) as

$$f\,(t, \eta(t,v)) = \overset{\circ}{f}(v) , \qquad v \in \mathbb{V} ,$$

which is well defined as long as $v \mapsto \eta(t,v)$ is a diffeomorphism. The characteristics do not intersect each other as long as

$$\frac{\partial}{\partial v}\eta(t,v) \neq 0 .$$

Given $v \in \mathbb{V}$, we denote $\eta_v(t) = \frac{\partial}{\partial v}\eta(t,v)$ and then we have the following problem for the ordinary differential equation

$$\begin{aligned} \frac{\mathrm{d}\,\eta_v(t)}{\mathrm{d}\,t} &= -\frac{\mathrm{d}}{\mathrm{d}\,v}\sigma(\eta(t))\, \eta_v(t)\, \overset{\circ}{f}{}^{\gamma}(v) \\ &\quad - \gamma\, \sigma(\eta(t))\, \overset{\circ}{f}{}^{\gamma-1}(v)\, \frac{\partial}{\partial v} \overset{\circ}{f}(v) , \\ \eta_v(0) &= 1 . \end{aligned} \tag{7.27}$$

We have the following estimation of its solution

$$\eta_v(t) > 1 - \int_0^t \left(\bar{M}_\sigma \parallel \overset{\circ}{f} \parallel_\infty^\gamma \eta_v(s) + M_\sigma \parallel \overset{\circ}{f} \parallel_\infty^{\gamma-1} \left\| \frac{\partial}{\partial v} \overset{\circ}{f} \right\|_\infty^\gamma \right) \mathrm{d}s$$

where

$$\bar{M}_\sigma = \sup_{v\in\mathbb{V}} \left| \frac{\mathrm{d}}{\mathrm{d}\,v} \sigma(v) \right|, \qquad \text{and} \qquad M_\sigma = \sup_{v\in\mathbb{V}} \left| \sigma(v) \right|.$$

It follows that $\eta_v(t) > 0$ in some interval $[0, \tau]$.

The derivative $\frac{\partial}{\partial v} f(t, v)$ satisfies the following formula

$$\frac{\partial}{\partial v} f(t, \eta(t, v)) \frac{\partial}{\partial v} \eta(t, v) = \frac{\partial}{\partial v} \overset{\circ}{f}(v),$$

which shows its boundedness for $t \in [0, \tau]$.

We take $p = 1$ and note that the arguments for $p > 1$ are similar. The derivative $\frac{\partial^2}{\partial v^2} f(t, v)$ satisfies the following formula

$$\frac{\partial^2}{\partial v^2} f(t, \eta(t, v)) \left(\eta_v(t, v)\right)^2 + \frac{\partial}{\partial v} f(t, \eta(t, v)) \frac{\partial^2}{\partial v^2} \eta(t, v) = \frac{\partial^2}{\partial v^2} \overset{\circ}{f}(v).$$

Keeping in mind $\overset{\circ}{f} \in L^1$, we have

$$\int_{\mathbb{V}} f(t, \eta(t, v))\, \mathrm{d}v = \int_{\mathbb{V}} f(t, \eta(t, v))\, \eta_v(t, v) \frac{1}{\eta_v(t, v)} \mathrm{d}v = \int_{\mathbb{V}} \overset{\circ}{f}(v)\, \mathrm{d}v.$$

Changing the variables $v \mapsto w' = \eta(t, v)$, yields

$$\frac{1}{\bar{M}_\eta} \int_{\mathbb{V}} f(t, w')\, \mathrm{d}w' \leq \int_{\mathbb{V}} \overset{\circ}{f}(v)\, \mathrm{d}v < \infty,$$

because

$$0 < \eta_v(t, v) \leq \bar{M}_\eta, \qquad \text{for} \quad t \in [0, \tau].$$

This ends the proof. ■ QED

The subsequent example; see Example 1 in Ref. [134], offers a noteworthy illustration.

Example 2 *Let*

- $\gamma \geq 1$,
- $\mathfrak{n} = \nu = 1$,
- $a(v) = \chi\,(0 \leq v \leq 1)$, *where* χ *is given by Eq.* (3.1).

Equations (7.17) *and* (7.22), *in this case, take the forms*

$$\frac{\partial}{\partial t} f(t, v) = \frac{1}{\varepsilon^2} \left(f^\gamma(t, v) \int_v^{v+\varepsilon} f(t, w)\, \mathrm{d}w - f(t, v) \int_{v-\varepsilon}^{v} f^\gamma(t, w)\, \mathrm{d}w \right). \tag{7.28}$$

and

$$\frac{\partial}{\partial t} f(t,v) = \frac{\gamma+1}{2} f^{\gamma}(t,v) \frac{\partial}{\partial v} f(t,v), \tag{7.29}$$

respectively.

The characteristics are given by

$$\eta(t,v) = v - t\frac{\gamma+1}{2} \overset{\circ}{f}{}^{\gamma}(v).$$

The solution defines the following formula

$$f\left(t, v - t\frac{\gamma+1}{2} \overset{\circ}{f}{}^{\gamma}(v)\right) = \overset{\circ}{f}(v), \tag{7.30}$$

or

$$f(t,v) = \overset{\circ}{f}\left(\tilde{v}(t,v)\right),$$

where $\tilde{v} = \tilde{v}(t,v)$ *is a solution of the following equation*

$$v = \tilde{v} - t\frac{\gamma+1}{2} \overset{\circ}{f}{}^{\gamma}(\tilde{v}). \tag{7.31}$$

Equation (7.31) *is solvable on* $[0,\tau]$, *with*

$$0 < \tau < \frac{2}{(\gamma+1)\gamma} \frac{1}{\| \overset{\circ}{f} \|_{\infty}^{\gamma-1} \left\| \frac{\partial}{\partial v} \overset{\circ}{f} \right\|_{\infty}}. \tag{7.32}$$

Taking $\gamma = 1$ *we obtain*

$$0 < \tau < \frac{1}{\left\| \frac{\partial}{\partial v} \overset{\circ}{f} \right\|_{\infty}}. \tag{7.33}$$

Following Ref. [134], We carry out an analysis using the Fourier transform. Let $\hat{f} = \mathcal{F}f$ denote the Fourier transform of $f \in \mathcal{S}'(\mathbb{V})$, cf. Ref. [4], with respect to the variable v,

$$\hat{f}(k) = \mathcal{F}f(k) = \frac{1}{(2\pi)^{\frac{\mathfrak{n}}{2}}} \int\limits_{\mathbb{V}} e^{-ik\cdot v} f(v) \mathrm{d}v, \quad k \in \mathbb{R}^{\mathfrak{n}}, \quad i = \sqrt{-1}.$$

Let $\mathbb{X}_{p,q}$ be the completion of the space that consists of functions f from $\mathcal{S}'(\mathbb{V})$ whose norm

$$\|f\|_{p,q} = \sup_{k\in\mathbb{R}^{\mathfrak{n}}} \left((1+|k|)^q \exp\left(p(1+|k|)\right) \left|\hat{f}(k)\right| \right),$$

is finite, where $p > 0$ and $q > \mathfrak{n}$; cf. the same setting in Ref. [206].

Additionally, let

$$\mathbb{X}_{p,q}^{(r)}(I) = \left\{ f = f(t) : \tilde{f}_r \in C\left(I; \mathbb{X}_{p,q}\right), \tilde{f}_r(t,v) \equiv \mathcal{F}^{-1} \exp(-rt|k|) \hat{f}(t,k) \right\}$$

be the space equipped with the norm

$$\|f\|_{p,q,I}^{(r)} = \sup_{t\in I} \|f(t)\|_{p-rt,q},$$

for $r \geq 0$ and a compact interval $I \subset \mathbb{R}$, where $C(I;\mathbb{X})$ denotes the space of $\mathbb{X}$–valued continuous functions on I.

The following lemma provides some elementary inequalities with straightforward proofs; see Lemma 1 in Ref. [134].

Lemma 2 *We have*

- *If* $q_1 > \mathfrak{n}$ *and* $q_2 > \mathfrak{n}$, *then*
$$\int_{\mathbb{R}^\mathfrak{n}} \frac{1}{(1+|k-\xi|)^{q_1}(1+|\xi|)^{q_2}}\,\mathrm{d}\xi \leq \frac{\mathfrak{c}}{(1+|k|)^{q_3}}\,, \qquad \forall\, k \in \mathbb{R}^\mathfrak{n}\,,$$
where $q_3 = \min\{q_1, q_2\}$ *and* $\mathfrak{c}$ *is a positive constant.*
- *If* $p > 0$, $r > 0$ *and* $t < \frac{p}{r}$, *then*
$$\int_0^t \exp\Big(-(p-rs)(1+|k|)\Big)\,\mathrm{d}s \leq \frac{\exp\Big(-(p-rt)(1+|k|)\Big)}{r\,(1+|k|)}\,,$$
for all $k \in \mathbb{R}^\mathfrak{n}$.

Moreover, we need the following basic lemma; see Lemma 2 in Ref. [134].

Lemma 3 *Let*

- $p > 0$, $q > \mathfrak{n}$ *and* $\delta > 0$,
- $\sigma \in \mathbb{X}_{p,q}$.

Then there exist $r > 0$ *and an interval* $I \subset [0,\infty)$ *such that* $0 \in I$ *and*
$$\|\varepsilon^{1-\nu}\int_I \sigma(v)\cdot\Big(g\,\frac{\partial}{\partial v}f\Big)\,\|^{(r)}_{p,q,I} \leq \delta\,\|g\|^{(r)}_{p,q,I}\,\|f\|^{(r)}_{p,q,I}\,, \tag{7.34}$$
for each $\varepsilon \in [0,1]$ *and* $\nu = 0$ *or* $\nu = 1$.

Proof. We follow Ref. [134]. Given $p > 0$, $q > \mathfrak{n}$ and $\delta > 0$. We note that the space $\mathbb{X}^{(r)}_{p,q}(I)$ is an algebra for any $r > 0$ and any interval I, that is, if both g and f are in $\mathbb{X}^{(r)}_{p,q}(I)$ then $gf \in \mathbb{X}^{(r)}_{p,q}(I)$. By Lemma 2, one can initially select $r > 0$ sufficiently large followed by I adequately small, but independent of ε, to fulfil Eq. (7.34). ■ QED

Using Lemma 3 yields the existence and uniqueness of the solutions for Eq. (7.26) on a small time interval that is independent of ε, for both $\nu = 0$ and $\nu = 1$ in the corresponding setting, and we have the following result; see Proposition 4 in Ref. [134].

Proposition 10 *Let*

- $\gamma \geq 1$, $p > 0$, $q > \mathfrak{n}$, $\delta > 0$,
- $\sigma \in \mathbb{X}_{p,q}$,
- $\mathring{f}$ *be such that* $\mathring{f} \in \mathbb{X}_{p,q}$.

Then there exist $r > 0$, $\tau > 0$ *and a unique solution* f *to Eq. (7.26) in* $\mathbb{X}^{(r)}_{p,q}\Big([0\,,\,\tau]\Big)$.

7.2.4 Asymptotic equivalence

We are going to impose the conditions that ensure both the local existence with uniqueness of the solution $f^\varepsilon = f^\varepsilon(t)$ for Eq. (7.19) as well as the local existence with uniqueness of the solution $f_B = f_B(t)$ for Eq. (7.22), both with $\nu = 0$ or $\nu = 1$. The aim is to show that the solutions are asymptotically equivalent on some interval $[0, \tau]$, with $\tau > 0$, that is the following statement holds

$$\sup_{t\in[0,\tau]} \left\| f^\varepsilon(t) - f_B(t) \right\| \le \mathfrak{c}\, \varepsilon^{2-\nu}\,, \tag{7.35}$$

where $\mathfrak{c}$ is a positive constant and $\| \,.\, \|$ is a suitable norm of a space in which both solutions exists.

We have the following asymptotic result; see Theorem 2 in Ref. [134].

Theorem 16 *Let*

- $\gamma \ge 1$, $\nu = 0$,
- $\overset{\circ}{f}$ *be a probability density in* $\mathbb{V}$ *such that* $\overset{\circ}{f} \in X^{(2)}$,
- $a \in X^{(2)}$ *satisfy Assumption* (8) *and*

$$\int_{\mathbb{V}} v^2\, a(v)\, dv \le \mathfrak{c}\,,$$

  where $\mathfrak{c}$ *is a positive constant,*

- $f^{(\varepsilon)} = f^{(\varepsilon)}(t)$, *defined on* $[0, \tau_K]$, *be a solution of Eq.* (7.19) *with the initial datum* $\overset{\circ}{f}$,
- $f_B = f_B(t)$, *defined on* $[0, \tau_B]$, *in* $X^{(1)}$ *be a solution to Eq.* (7.22) *with the same initial data* $\overset{\circ}{f}$.

Then f^ε *and* f_B *are asymptotically equivalent, i.e.*

$$\sup_{t\in[0,\tau]} \left\| f^{(\varepsilon)}(t) - f_B(t) \right\|_\infty \le \mathfrak{c}\, \varepsilon^2\,,$$

where $\tau = \min\{\tau_K, \tau_B\}$ *and* $\mathfrak{c}$ *is a positive constant.*

Proof. We follow Ref. [134]. Assume that $f^{(\varepsilon)} = f^{(\varepsilon)}(t)$ and $f_B = f_B(t)$ are the solutions to Eq. (7.19) and Eq. (7.22), respectively, both given on interval $[0, \tau]$, for some $\tau > 0$. We can assume, for simplicity, that $\gamma \ge 1$ is an integer. It follows

$$\begin{aligned}
&\tfrac{\partial}{\partial t}\left(f^{(\varepsilon)}(t,v) - f_B(t,v)\right) \; = \\
&= \; \left(f^{(\varepsilon)}(t,v)\right)^\gamma \int_{\mathbb{V}} a(w') \left(f^{(\varepsilon)}(t, v + \varepsilon w') - f^{(\varepsilon)}(t,v)\right) dw' \\
&- \; f^{(\varepsilon)}(t,v) \int_{\mathbb{V}} a(w') \left(\left(f^{(\varepsilon)}(t, v - \varepsilon w')\right)^\gamma - \left(f^{(\varepsilon)}(t,v)\right)^\gamma\right) dw' \\
&- \; \varepsilon\,(1+\gamma)\,\left(f_B(t,v)\right)^\gamma\, B_1 \cdot \tfrac{\partial}{\partial v} f_B(t,v)\,,
\end{aligned}$$

Taylor's expansion yields

$$\frac{\partial}{\partial t}\Big(f^{(\varepsilon)}(t,v) - f_B(t,v)\Big) =$$

$$= \varepsilon\,(1+\gamma)\,\left(f_B(t,v)\right)^{\gamma}\,B_1\cdot\frac{\partial}{\partial v}\Big(f^{(\varepsilon)}(t,v) - f_B(t,v)\Big) +$$

$$+ \varepsilon\,(1+\gamma)\Big(\left(f^{(\varepsilon)}(t,v)\right)^{\gamma} - \left(f_B(t,v)\right)^{\gamma}\Big)B_1\cdot\frac{\partial}{\partial v}f^{(\varepsilon)}(t,v) +$$

$$+ \frac{\varepsilon^2}{2}\,\left(f^{(\varepsilon)}(t,v)\right)^{\gamma}\sum_{i,j=1}^{n}\int_{\mathbb{V}} a(w')\,w'_i\,w'_j \times$$

$$\times \frac{\partial^2}{\partial v_i\,\partial v_j}\,f^{(\varepsilon)}\Big(t, v+\theta_1(\varepsilon w' - v)\Big)\,\mathrm{d}w'$$

$$- \frac{\varepsilon^2}{2}\,f^{(\varepsilon)}(t,v)\sum_{i,j=1}^{n}\int_{\mathbb{V}} a(w')\,w'_i\,w'_j$$

$$\times \frac{\partial^2}{\partial v_i\,\partial v_j}\Big(f^{(\varepsilon)}\Big(t, v-\theta_2(\varepsilon w' + v)\Big)\Big)^{\gamma}\,\mathrm{d}w' ,$$

where $\theta_1, \theta_2 \in [0,1]$.

Keeping in mind the characteristics defined in Section 7.2.3 we represent the difference

$$f^{(\varepsilon)}(t,v) - f_B(t,v)$$

by the following equation

$$\frac{\partial}{\partial t}\Big(f^{(\varepsilon)}(t,v) - f_B(t,v)\Big)^{\#} =$$

$$= \varepsilon\,(1+\gamma)\Big(\left(f^{(\varepsilon)}(t,v)\right)^{\gamma} - \left(f_B(t,v)\right)^{\gamma}\Big)^{\#}\Big(B_1\cdot\frac{\partial}{\partial v}f^{(\varepsilon)}(t,v)\Big)^{\#} +$$

$$+ \Bigg(\frac{\varepsilon^2}{2}\,\left(f^{(\varepsilon)}(t,v)\right)^{\gamma}\sum_{i,j=1}^{d}\int_{\mathbb{V}} a(w')\,w'_i\,w'_j \times$$

$$\times \frac{\partial^2}{\partial v_i\,\partial v_j}\,f^{(\varepsilon)}\Big(t, v+\theta_1(\varepsilon w' - v)\Big)\,\mathrm{d}w'$$

$$- \frac{\varepsilon^2}{2}\,f^{(\varepsilon)}(t,v)\sum_{i,j=1}^{d}\int_{\mathbb{V}} a(w')\,w'_i\,w'_j \times$$

$$\times \frac{\partial^2}{\partial v_i\,\partial v_j}\Big(f^{(\varepsilon)}\Big(t, v-\theta_2(\varepsilon w' + v)\Big)\Big)^{\gamma}\,\mathrm{d}w'\Bigg)^{\#} ,$$

where

$$f^{\#}(t,v) = f\left(t, \eta(t,v)\right) ,$$

see the proof of Theorem 15.

Thus, under the assumptions of Theorem 16, by Proposition 9, the statement of Theorem 16 is proved. ■ QED

By Proposition 7 we know that there is a global solution for $\gamma = 1$ in L^1–setting. This enables us to propose a hypothesis regarding the weak asymptotic result on the interval $[0, \tau]$ as in Theorem 15. In such a case, the derivatives of the solution exist in each time interval, but the norms of the derivatives are not independent of ε. We can take advantage of the estimations in L^1 in a one–dimensional setting using test functions. Again, the Taylor expansion and the equation along characteristics can be employed together with the fact that the second–order terms vanish. This provides a foundation for Conjecture 1 in Section 14 that was stated in Ref. [134].

We now return to the Fourier analysis discussed in Section 7.2.3. We take $\nu = 1$ and denote

$$\delta(t, v) = f^{(\varepsilon)}(t, v) - f_B(t, v) .$$

For the sake of simplicity, we omit the indication of dependence on ε. It follows

$$\begin{aligned} \frac{\partial}{\partial t}\delta(t, v) &= \frac{1}{\varepsilon^{\mathfrak{n}+1}} \Bigg(\left(f^{(\varepsilon)}(t, v)\right)^\gamma \int_{\mathbb{V}} a\left(\tfrac{w-v}{\varepsilon}\right) f^{(\varepsilon)}(t, w)\, \mathrm{d}w \\ &- f^{(\varepsilon)}(t, v) \int_{\mathbb{V}} a\left(\tfrac{v-w}{\varepsilon}\right) \left(f^{(\varepsilon)}(t, w)\right)^\gamma \mathrm{d}w \Bigg) \\ &- 2\left(f_B(t, v)\right)^\gamma B_1 \cdot \frac{\partial}{\partial v} f_B(t, v) . \end{aligned}$$

Taking $\gamma = 1$ yields

$$\frac{\partial}{\partial t}\delta(t, v) = f^{(\varepsilon)}(t, v)\, L_\varepsilon f^{(\varepsilon)}(t, v) - 2\, f_B(t, v)\, B_1 \cdot \frac{\partial}{\partial v} f_B(t, v) , \tag{7.36}$$

where L_ε is a linear operator given by

$$\begin{aligned} L_\varepsilon f(v) &= \frac{1}{\varepsilon^{\mathfrak{n}+1}} \int_{\mathbb{V}} \left(a\left(\tfrac{w-v}{\varepsilon}\right) - a\left(\tfrac{v-w}{\varepsilon}\right)\right) f(w)\, \mathrm{d}w = \\ &= \frac{1}{\varepsilon} \int_{\mathbb{V}} \Big(a(w') - a(-w') \Big) f(v + \varepsilon w')\, \mathrm{d}w' . \end{aligned} \tag{7.37}$$

We observe that for $\mathbb{V} = \mathbb{T}^{\mathfrak{n}}$, that is, when $\mathbb{V}$ is the $\mathfrak{n}$–dimensional torus, such that $|\mathbb{V}| = 1$, the function $f_{\mathrm{eq}} \equiv 1$ is an equilibrium solution, that is not necessarily unique. It means that

$$\mathcal{R}_\varepsilon(f_{\mathrm{eq}}) = 0 \qquad \forall\, \varepsilon > 0 .$$

Then L_ε is the Fréchet derivative of the operator $f \to \mathcal{R}_\varepsilon(f)$ in $f = f_{\mathrm{eq}} = 1$, that is, in other words, the linearization of $\mathcal{R}_\varepsilon$ around the equilibrium solution f_{eq}.
If $a \in L^1$ we have

$$\begin{aligned} &\int_{\mathbb{V}} \exp\Big(i\, \varepsilon\, k \cdot w'\Big) \Big(a(w') - a(-w')\Big)\, \mathrm{d}w' = 2\, i \int_{\mathbb{V}} \sin\Big(\varepsilon\, k \cdot w'\Big)\, a(w')\, \mathrm{d}w' = \\ &=: \; i\, \lambda_{k,\varepsilon} , \end{aligned}$$

for all $k \in \mathbb{R}^{\mathfrak{n}}$ and all $\varepsilon > 0$. Moreover,

$$\widehat{L_\varepsilon f}(k) = i\, \frac{\lambda_{k,\varepsilon}}{\varepsilon}\, \hat{f}(k) ,$$

for all $k \in \mathbb{R}^{\mathfrak{n}}$ and all $\varepsilon > 0$.

Consequently, the Fourier transform of the right side of Eq. (7.36) is given by

$$\mathcal{F}\Big(f^{(\varepsilon)}\, L_\varepsilon f^{(\varepsilon)} - 2 f_B\, B_1 \cdot \tfrac{\partial}{\partial v} f_B \Big)(t,k) \; =$$

$$= \frac{i}{(2\,\pi)^{\frac{n}{2}}} \left(\frac{1}{\varepsilon} \int_{\mathbb{R}^n} \hat{f}^{(\varepsilon)}(t, k-\xi)\, \lambda_{\xi,\varepsilon}\, \hat{f}^{(\varepsilon)}(t,\xi)\, \mathrm{d}\xi \right. \tag{7.38}$$

$$\left. - \; 2 \int_{\mathbb{R}^n} f_B(t, k-\xi)\, B_1 \cdot \xi\, f_B(t,\xi)\, \mathrm{d}\xi \right) .$$

Keeping in mind that initially

$$\delta(0,v) \,=\, 0 \,, \qquad \forall \quad v \in \mathbb{V} \,,$$

the equation for δ reads

$$\hat{\delta}(t,k) \;=$$

$$= \frac{i}{(2\,\pi)^{\frac{n}{2}}} \left(\int_0^t \left(\int_{\mathbb{R}^n} \left(\left(\frac{\lambda_{\xi,\varepsilon}}{\varepsilon} - 2\, B_1 \cdot \xi \right) \hat{f}_B(t, k-\xi)\, \hat{f}_B(t,\xi) \, + \right.\right.\right.$$

$$+ \;\; \frac{\lambda_{\xi,\varepsilon}}{\varepsilon} \Big(\hat{f}_B(s, k-\xi) \hat{\Delta}(s,\xi) \, + \tag{7.39}$$

$$\left.\left.\left. + \;\; \hat{\Delta}(s, k-\xi) \hat{f}_B(s,\xi) + \hat{\Delta}(s,k-\xi)\hat{\delta}(s,\xi) \Big)\right) \mathrm{d}\xi \right) \mathrm{d}s \right) .$$

Using the elementary inequalities,

$$|\sin x| \,\le\, |x| \,, \qquad |x - \sin x| \,\le\, 2\,|x| \,, \qquad \forall\, x \in \mathbb{R} \,, \tag{7.40}$$

we get

$$\left| \frac{\lambda_{\xi,\varepsilon}}{\varepsilon} \right| \,=\, 2 \left| \int_{\mathbb{V}} \frac{\sin(\varepsilon\, \xi \cdot w)}{\varepsilon} a(w)\, \mathrm{d}w \right| \le 2\,|\xi|\,|B_1| \,, \tag{7.41}$$

and

$$\left| \frac{\lambda_{\xi,\varepsilon}}{\varepsilon} - 2\, B_1 \cdot \xi \right| \,=\, 2 \left| \int_{\mathbb{V}} \left(\frac{\sin(\varepsilon\, \xi \cdot w)}{\varepsilon} - \xi \cdot w \right) a(w)\, \mathrm{d}w \right| \le 4\,|\xi|\,|B_1| \,. \tag{7.42}$$

By the inequality

$$|x - \sin x| \,\le\, \frac{x^2}{2} \qquad \forall\, x \in \mathbb{R} \,,$$

we obtain

$$\left| \frac{\lambda_{\xi,\varepsilon}}{\varepsilon} - 2\, B_1 \cdot \xi \right| \,\le\, \varepsilon\, |\xi|^2 \int_{\mathbb{V}} w^2 a(w)\, \mathrm{d}w \,. \tag{7.43}$$

The theorem regarding asymptotic equivalence can now be formulated; see Theorem 3 in Ref. [134].

Theorem 17 *Let*

- $\gamma \geq 1$, $p > 0$, $q > \mathfrak{n}$;
- $a \in \mathbb{X}_{p,q}$ *satisfy Assumption* (8);
- $\overset{\circ}{f} \in \mathbb{X}_{p,q}$.

Then there exist $r > 0$, $\tau > 0$ *and unique solutions* $f^{(\varepsilon)}$ *and* f_B *to Eq.* (7.19) *and Eq.* (7.26), *respectively, both in* $\mathbb{X}_{p,q}^{(r)}\Big([0\,,\,\tau]\Big)$ *and with the same initial datum* $\overset{\circ}{f}$. *If additionally,*

$$\int\limits_{\mathbb{V}} v^2\, a(v)\, \mathrm{d}v \; < \; \infty\,.$$

then f^{ε} *and* f_B *are asymptotically equivalent, that is,*

$$\left\| f^{(\varepsilon)} - f_B \right\|_{p,q-1,[0,\tau]}^{(r)} \; \leq \; \mathfrak{c}\,\varepsilon\,, \tag{7.44}$$

where $\mathfrak{c}$ *is a positive constant.*

Proof. We follow Ref. [134]. A similar approach can be used to study the case in which $\gamma > 1$, like the situation when $\gamma = 1$. If the assumptions of Theorem 17 are satisfied, by Proposition 10, the result concerning existence and uniqueness of solutions for Eq. (7.26) is ensured. Next, utilizing Eqs. (7.41) and (7.42), we are able to apply methods like those used in the demonstration of Lemma 3. This allows us first select $r > 0$ to be sufficiently large and then τ small enough such that the operator defined by the right–hand side of Eq. (7.39) is contracting in the space $\mathbb{X}_{p,q}^{(r)}\Big([0,\tau]\Big)$. Thus, it follows that there is a unique solution δ in $\mathbb{X}_{p,q}^{(r)}\Big([0,\tau]\Big)$. Then we conclude that the result concerning the existence and uniqueness of solutions for Eq. (7.19) in $\mathbb{X}_{p,q}^{(r)}\Big([0,\tau]\Big)$ holds. Finally, by Eq. (7.43), we obtain Eq. (7.44) which ends the proof. ■ QED

Following Ref. [134] we are able to present an interesting example that demonstrates a type of singular behaviour.

Example 3 *Assume that*

$$a(v) \; = \; \chi\,(0 \leq v \leq 1)\;,$$

where χ *is given by Eq.* (3.1) *and* $\mathbb{V}$ *is the* 1*–dimensional torus* $\mathbb{V} = \mathbb{T}^1$. *Thus, the operator* L_ε, *defined by Eq.* (7.37), *has the form*

$$L_\varepsilon f(v) \; = \; \frac{1}{\varepsilon^2}\left(\int\limits_{v}^{v+\varepsilon} f(w)\,\mathrm{d}w - \int\limits_{v-\varepsilon}^{v} f(w)\,\mathrm{d}w \right). \tag{7.45}$$

We can see that the unique solution of the equation

$$\frac{\partial}{\partial t} f \; = \; L_\varepsilon f\,, \tag{7.46}$$

corresponding to the initial datum

$$\overset{\circ}{f}(v) = \varepsilon \sin \frac{\pi v}{\varepsilon}, \tag{7.47}$$

is defined by the function

$$f(t, v) = \varepsilon \sin \frac{\pi v}{\varepsilon} \cosh \frac{4t}{\pi \varepsilon} - \varepsilon \cos \frac{\pi v}{\varepsilon} \sinh \frac{4t}{\pi \varepsilon}. \tag{7.48}$$

We can see that $|f(t, v)|$ *can become arbitrarily large, although the initial data* $\overset{\circ}{f}(v)$ *approach* 0 *as* $\varepsilon \to 0$.

The importance of asymmetric interaction is vital in the domain of mathematical economics. In general, the effect one agent has on another is not the same as the effect the latter agent has on the first. The fundamental lack of symmetry in mathematical structures introduces additional complexity to mathematical frameworks. In fact, the complexity of mathematical theories regarding asymmetric structures is markedly greater than that associated with symmetric structures such as those considered in Refs. [136–139, 170]. We have formally demonstrated that the solutions may exhibit singular behaviour. The approach presented herein can be regarded as an initial investigation into mathematical structures and may serve as a foundation for future research. The primary unresolved question is related to Conjecture 1 in Section 14. There is an expectation that mathematical theory can provide a more precise description of processes in the life and social sciences.

Chapter 8

Equilibrium solutions

As before, $|\mathbb{U}|$ denotes the Lebesgue measure of the measurable set $\mathbb{U} \subset \mathbb{V}$.

This chapter addresses the classification of equilibrium solutions, i.e. our objective is here to determine the complete class of equilibrium solutions, that is, all probability densities that satisfy the following equation

$$Q[f] = 0\,, \tag{8.1}$$

where Q is the nonlinear operator defined by the right–hand side of Eq. (2.19).

In Ref. [170], this issue was solved for the set $\mathbb{V}$ that is compact and the interaction rate a that is distinct from 0. If Assumption 4 holds, the class includes all functions that are constant on their supports. In fact, we have; see Proposition 3 in Ref. [170],

Theorem 18 *Let*

- $\mathbb{V}$ *be compact,*
- $\gamma > 0$ *and* $\gamma \neq 1$,
- *Assumption 4 be satisfied.*

Then, a probability density f on $\mathbb{V}$ satisfies Eq. (8.1) *if and only if there exists a measurable set $\mathbb{U} \subset \mathbb{V}$ such that*

$$f(v) = \frac{1}{|\mathbb{U}|}\,\chi\,(v \in \mathbb{U})\;,$$

where χ is given by Eq. (3.1).

Proof. See Ref. [170], the proof of Proposition 3. ■ QED

None of the equilibrium solutions, in this case, can be asymptotically stable.

Let $\gamma > 1$ and $\mathbb{V} = \mathbb{R}^n$. Writing the set $|\mathbb{U}|$, we always assume that this set is a measurable subset of $\mathbb{V} = \mathbb{R}^n$.

A more general class of interaction rates was considered in Ref. [139] that leads to a much more complex class of equilibrium solutions. The equilibrium solutions are functions that are constant in a countable family of subsets of their supports if these subsets are distant.

We denote

$$\operatorname{supp} a \;=\; \Big\{(v,w) \in \mathbb{V}\times\mathbb{V}\;:\; a(v,w) > 0\Big\}\,.$$

We assume; see Ref. [139],

Assumption 9 *Let*

- *the function $a\;:\;\mathbb{V}\times\mathbb{V} \mapsto \mathbb{R}_+$, $a \in L^\infty(\mathbb{V}\times\mathbb{V})$ be symmetric, i.e., Eq.* (2.7) *be satisfied,*

DOI: 10.1201/9781003537595-8

- *given a measurable set* $\mathbb{U} \subset \mathbb{V}$ *such that* $|\mathbb{U}| > 0$ *and*

$$|(\mathbb{U} \times \mathbb{U}) \cap \operatorname{supp} a| > 0\,.$$

By Assumption 9 we obtain

$$\Big| \operatorname{supp} a \Big| > 0\,.$$

Following Ref. [139] we introduce the concept of a–separable sets.

Definition 1 *Two sets* $\mathbb{U}_1, \mathbb{U}_2 \subset \mathbb{V}$ *are* a*–separable if*

$$\Big| (\mathbb{U}_1 \times \mathbb{U}_2) \cap \operatorname{supp} a \Big| = 0\,.$$

The meaning of Assumption 9 is that a positive measure set $\mathbb{U}$ is not a–separable from itself. Moreover, under Assumption 9, if $\mathbb{U}_1, \mathbb{U}_2$ are a–separable, then $|\mathbb{U}_1 \cap \mathbb{U}_2| = 0$.

Following Ref. [139] we introduce the following concepts.

Definition 2 *We define*

- $\mathbb{U} \subset \mathbb{V}$ *is called an a atom if there is no partition* $\mathbb{U} = \mathbb{U}_1 \cup \mathbb{U}_2$ *such that* $|\mathbb{U}_1| > 0$, $|\mathbb{U}_2| > 0$, $\mathbb{U}_1 \cap \mathbb{U}_2 = \emptyset$ *and* $\mathbb{U}_1, \mathbb{U}_2$ *are* a*–separable,*
- $\mathbb{U}$ *is a maximal set with this property, which means that it is not contained in another set that has this property,*
- *A countable partition* $\{\mathbb{U}_1, \mathbb{U}_2, \dots\}$ *of a set* $\mathbb{U}$ *is called* a*–partition of* $\mathbb{U}$ *if all sets* $\mathbb{U}_i$*,* $i = 1, 2, \dots,$ *are mutually distinct* a*–atoms, of positive Lebesgue measure, and* $\mathbb{U}_i$*,* $\mathbb{U}_j$ *are* a*–separable for* $i \neq j$*.*

We have the following result; see Ref. [139],

Lemma 4 *Let*

- *Assumption 9 be fulfilled,*
- $|\mathbb{U}| > 0$.

Then there exists a*–partition of* $\mathbb{U}$*. The partition is unique with respect to sets of measure zero.*

Proof. See the proof of Lemma 2.1 in Ref. [139]. ■ QED

Any function a that fulfils Assumption 6, with some $\varepsilon > 0$, also meets the requirements of Assumption 9. This consequently leads to a range of significant examples that could be relevant for the modelling of biological and social processes.

We have; see Ref. [139],

Theorem 19 *Let*

- *Assumption 9 be satisfied;*
- $f \in L^\infty \cap L^1$ *be a probability density on* $\mathbb{V} = \mathbb{R}^n$.

Then f *is a equilibrium solution if and only if*

$$f(v) = \sum_i c_i \chi(v \in \mathbb{U}_i) ,$$

where $\{\mathbb{U}_i\}$ *is a–partition of the set* $\mathbb{U} = \{v : f(v) > 0\}$,

$$c_i > 0 , \qquad \sum_i c_i |\mathbb{U}_i| = 1 , \qquad \sup_i c_i < \infty ,$$

and χ *is given by Eq.* (3.1)

Proof. See the proof of Theorem 2.3 in Ref. [139]. ■ QED

The equilibrium solutions derived from Theorem 19 do not exhibit asymptotic stability.

The following corollary on far–distant patches refers to a which corresponds to short–distance interactions; see Corollary 2.4 in Ref. [139].

Proposition 11 *Let*

- a *be nonnegative,* $a \in L^\infty(\mathbb{V} \times \mathbb{V})$, *and satisfy Assumption 6 with some* $\varepsilon > 0$;
- $f \in L^\infty \cap L^1$ *be a probability density on* $\mathbb{V} = \mathbb{R}^n$.

Then f *is a equilibrium solution if and only if*

$$f(v) = \sum_{i \in \mathbb{N}} \mathfrak{c}_i \chi(v \in \mathbb{U}_i) , \tag{8.2}$$

where $\mathfrak{c}_i$ *are positive real numbers,* $\mathbb{U}_i$ *are measurable sets of positive measure whose mutual distances are not smaller than* ε, *and*

$$\sum_{i \in \mathbb{N}} c_i |\mathbb{U}_i| = 1 . \tag{8.3}$$

The interpretation of Proposition 11 could be as follows. The rate of interactions that occur within the short range defined by ε results in the support of any equilibrium solution that can be in patches that are distant from each other determined by the distance of interaction defined by ε. The constant values on each patch are constrained only by Eq. (8.3).

Chapter 9

Microscopic scale

9.1 Microscopic model

In the previous chapters, we considered the description on the mesoscopic scale in which the function defining the state of systems refers to one test–agent. It is possible to propose an alternative approach in which one can consider an entirely microscopic description that refers to the states of all agents involved in the process.

Such a fully microscopic description can be deterministic, as in the case of Newton's laws, or stochastic. In both cases, the literature is enormous. Here we consider a stochastic description of all agents involved in the process; see Refs. [24,130,131] and references therein.

We propose a general linear evolution equation that is defined by a generator—a linear operator that can be related to a Markov jump process describing changes of states as a result of interactions; see Ref. [130]. The equation defines the evolution of the probability density related to the states of number N of agents that belong to some sub–populations of the system referred to as the general population. Any agent n, $n \in \{1, \dots, N\}$, can be characterized by its microscopic state,

$$v_n \in \mathbb{V}.$$

The variable v_n, related to the agent n, can have various meanings, as we already discussed in Chapters 1 and 2. Specifically, a discrete component in the case [**CD**] in Section 2.2 has the ability to contain data on a subset, that is, sub–population of agents of which the agent is a part—see [24, 130, 131].

In the general setting $(\mathbb{V}, \mathfrak{B}, \mu)$ is a σ–finite measure space. In case [**CD**] in Section 2.2, $\mathbb{V}$ is a product of a discrete set and a Lebesgue measurable subset. The measure μ is the product of the counting measure and the Lebesgue measure. We can also consider the discrete–discrete case, in which the measure μ is a product of the counting measures. In this chapter, as before in this monograph, we adopt the Lebesgue measure notation for simplicity, and we use Eq. (2.17).

The state v_{n_1} of each agent n_1, where $n_1 \in \{1, 2, \dots, N\}$, changes stochastically at random times with the change that exhibits a jumping nature. Following Refs. [24, 130], we permit the following changes,

- without any interaction with other agents,
- due to interaction with the agent n_2, $n_2 \neq n_1$,
- due to interaction with the agents n_2 and n_3, $n_i \neq n_j$, if $i \neq j$, $i, j = 1, 2, 3$,
- ...,
- due to interaction with n_2, n_3, ..., n_M agents, $n_i \neq n_j$, if $i \neq j$, $i, j = 1, 2, \dots, M$, where M is an integer $1 \leq M \leq N$.

DOI: 10.1201/9781003537595-9

Still following Refs. [24, 130], we define all acceptable interactions between the agent n_1 and the other $m-1$ agents, where m is given $m = 1, \ldots, M$ and $M \leq N$. The functions of Assumptions 10 and 11 define these interactions.

Assumption 10 *The rate of interaction between the agent with state v_{n_1} and the agents with states v_{n_2}, ..., v_{n_m} is defined by the measurable function $a^{[m]} = a^{[m]}(v_{n_1}, \ldots, v_{n_m})$, such that*

$$0 \leq a^{[m]}(v_{n_1}, v_{n_2}, \ldots, v_{n_m}) \leq a_+^{[m]}, \tag{9.1}$$

for all $v_{n_1}, v_{n_2}, \ldots, v_{n_m} \in \mathbb{V}$, where $a_+^{[m]} < \infty$ is a constant.

The bound condition stated in Eq. (9.1) could be quite restrictive in certain applications. However, it notably simplifies the mathematical framework.

Assumption 11 *The jump to state v for agent n_1 which currently posses state v_{n_1}, as a result of interaction with agents n_2,...,n_m with states v_{n_2},...,v_{n_m}, respectively, is defined by the measurable function $A^{[m]} = A^{[m]}(v; v_{n_1}, \ldots, v_{n_m}) \geq 0$, where*

$$\int_{\mathbb{V}} A^{[m]}\Big(v\,;\, v_{n_1}, v_{n_2}, \ldots, v_{n_m}\Big)\, dv = 1\,, \tag{9.2}$$

for all $v_{n_1}, v_{n_2}, \ldots, v_{n_m} \in \mathbb{V}$.

Equation (9.2) characterizes the conservative nature of the description, highlighting its probabilistic aspect. For certain applications, it can be possible to relax the conservative nature if necessary. However, this can lead to additional mathematical difficulties.

The stochastic model on the microscopic scale is completely defined by selecting the functions $a^{[m]}$ and $A^{[m]}$.

Let $L^1_{(N)}$ be the space equipped with the norm

$$\|f\|_{L^1_{(N)}} = \Big(\int_{\mathbb{V}}\Big)^N \Big|f(v_1, \ldots, v_N)\Big|\, dv_1 \ldots dv_N\,.$$

We use Convention Eq. (2.11) and if the value of N is 1, we can express it by L^1, as before. Given N, M, and $a^{[m]}$, $A^{[m]}$, for $m = 1, \ldots, M$, we propose the stochastic model that is defined by the Markov jump process of N agents; see Ref. [24], through the following generator Λ that acts on probability densities rather than on general measures

$$\Lambda = \Lambda^+ - \Lambda^- = \sum_{m=1}^{M} \Big(\Lambda^{[m]+} - \Lambda^{[m]-}\Big),$$

and

$$\begin{aligned}
&\Lambda^{[m]+} f\big(t, v_1, v_2, ..., v_N\big) = \\
&= \mathfrak{c}_{N,m} \sum_{\substack{1 \leq n_1, \ldots, n_m \leq N \\ n_i \neq n_j\ \forall i \neq j}} \int_{\mathbb{V}} A^{[m]}\big(v_{n_1}; w, v_{n_2}, \ldots, v_{n_m}\big) a^{[m]}\big(w, v_{n_2}, \ldots, v_{n_m}\big) \times \\
&\times\ f\big(t, v_1, \ldots, v_{n_1-1}, w, v_{n_1+1}, \ldots, v_N\big)\, dw\,,
\end{aligned}$$

and

$$\Lambda^{[m]-} f(t, v_1, v_2, ..., v_N) =$$

$$= \mathfrak{c}_{N,m} \sum_{\substack{1 \le n_1,\dots,n_m \le N \\ n_i \ne n_j \ \forall i \ne j}} a^{[m]}(v_{n_1}, v_{n_2}, \dots, v_{n_m}) f(t, v_1, \dots, v_N),$$

on $\mathbb{V}^N$, where $\mathfrak{c}_{N,m} = \frac{1}{(m-1)!\binom{N}{m-1}}$ are the normalizing constants. The linear operator Λ has a standard structure—it is the difference between the gain term Λ^+ and the loss term Λ^- that describe the gains and losses as the results of interactions.

When the system initially is distributed by an initial probability density $\overset{\circ}{f} \in L^1_{(N)}$ then the behaviour in time in the space $L^1_{(N)}$ is determined by the following linear Cauchy problem,

$$\begin{aligned} \frac{\partial}{\partial t} f &= \Lambda f, \\ f\Big|_{t=0} &= \overset{\circ}{f}. \end{aligned} \tag{9.3}$$

Equation (9.3) can be treated as a modified Liouville equation. In fact, this equation serves a purpose similar to the classical Liouville equation within the kinetic theory of N particles; see Ref. [54]. The generator $\Lambda = \Lambda^+ - \Lambda_-$ can be clearly understood as follows; see Ref. [130],

- The gain term Λ^+ is expressed as a sum over $m = 1, 2, \dots, M$, and includes terms that describe the jump of agent n_1 from state v to v_{n_1} due to interactions with agents $n_2, \dots, n_m$ with states $v_{n_2}, \dots, v_{n_m}$, respectively, where $2 \le m \le M$. Furthermore, it incorporates a term for $m = 1$ representing the jump of agent n_1 from state v to u_{n_1} without any interactions.
- Similarly, the loss term Λ^- is the sum of terms that define the jump from state v_{n_1} of agent n_1 to another state due to interaction with agents n_2, ..., n_m with states v_2, ..., v_m, respectively, for $2 \le m \le M$ or without any interaction for $m = 1$.

Keeping in mind Assumptions 10 and 11, it is evident that the operator Λ is bounded in $L^1_{(N)}$. This certainly simplifies the mathematical theory and it follows that the Cauchy problem has the unique solution

$$f(t) = \exp(t\,\Lambda)\ \overset{\circ}{f}, \tag{9.4}$$

in $L^1_{(N)}$ for all $t \ge 0$. Standard arguments demonstrate that the solution remains nonnegative when the initial data $\overset{\circ}{f}$ are nonnegative and that the $L^1_{(N)}$–norm is preserved

$$\| f(t) \|_{L^1_{(N)}} = \| \overset{\circ}{f} \|_{L^1_{(N)}} = 1, \qquad \text{for } t > 0. \tag{9.5}$$

Thus, $\{\exp(t\Lambda)\}_{t\ge 0}$ defines a uniformly continuous semigroup of Markov operators, that is, a uniformly continuous stochastic semigroup, see Ref. [148].

We want to consider the situation in which all functions are symmetric when their variables are permuted

$$f(v_1, \dots, v_N) = f(v_{r_1}, \dots, v_{r_N}), \tag{9.6}$$

for all v_1,...,v_N in $\mathbb{V}$ and for any permutation $\{r_1, \dots, r_N\}$ of the set $\{1, \dots, N\}$. We introduce the s–agent marginal density, $1 \le s < N$,

$$f^s(v_1, \dots, v_s) = \Big(\int_{\mathbb{V}}\Big)^{N-s} f(v_1, \dots, v_N)\, dv_{s+1} \dots dv_N, \tag{9.7}$$

with the convention $f^{N,N} = f^N$ and $f^{N,s'} \equiv 0$ if $s' > N$ and Convention defined by Eq. (2.11).

Let the solution (that is, Eq. (9.4)) to the Cauchy problem Eq. (9.3) be now denoted by $f^N = f^N(t)$. Then the corresponding $f^{N,s}$ satisfy the finite hierarchy of equations; see Ref. [130],

$$\frac{\partial}{\partial t} f^{N,s} = \mathcal{A}^s_{N,M}\Big(f^{N,s}, f^{N,s+1}, \dots, f^{N,s+M-1}\Big), \tag{9.8}$$

for $s = 1, 2, \dots, N$, where the linear operators $\mathcal{A}^s_{N,M}$ are given by

$$\begin{aligned} \mathcal{A}^s_{N,M}\Big(f^{N,s}, \dots, f^{N,s+M-1}\Big) &= \\ &= \sum_{m=1}^{M} \frac{(N-m+1)!}{N!} \mathcal{A}^{s,m}_{N,M}\Big(f^{N,s}, \dots, f^{N,s+m-1}\Big), \end{aligned} \tag{9.9}$$

and

$$\begin{aligned} &\mathcal{A}^{s,m}_{N,M}\Big(f^{N,s}, \dots, f^{N,s+m-1}\Big)(v_1, \dots, v_s) = \\ &= \sum_{k=1}^{m} \frac{(N-s)!}{(N-s-m+k)!} \sum_{\substack{1 \le n_1, \dots, n_k \le s \\ n_i \neq n_j \ \forall i \neq j}} \sum_{\{\}} \Big(\int_{\mathbb{V}}\Big)^{m-k} \\ &\Big(\int_{\mathbb{V}} A^{[m]}\left(v_{n_1}; w, \left\{v_{n_2}, \dots, v_{n_k}, v_{s+1}, \dots v_{s+m-k}\right\}\right) \times \\ &\times\ a^{[m]}\left(w, \left\{v_{n_2}, \dots, v_{n_k}, v_{s+1}, \dots, v_{s+m-k}\right\}\right) \times \\ &\times\ f^{N,s+m-k}\left(v_1, \dots, v_{n_1-1}, w, v_{n_1+1}, \dots, v_s, \dots, v_{s+m-k}\right) dw \\ &-\ a^{[m]}\left(v_{n_1}, \left\{v_{n_2}, \dots, v_{n_k}, v_{s+1}, \dots, v_{s+m-k}\right\}\right) \times \\ &\times\ f^{N,s+m-k}\left(v_1, \dots, v_{s+m-k}\right)\Big) dv_{s+1} \dots dv_{s+m-k}, \end{aligned} \tag{9.10}$$

and we apply the convention that $\sum_{\{\}}$ means the sum over all permutation of variables within $\{\}$ as well as if $m-k<1$, then

$$\sum_{\{\}} \Big(\int_{\mathbb{V}}\Big)^{m-k},$$

$\{v_{s+1}, \dots v_{s+m-k}\}$, and $dv_{s+1} \dots dv_{s+m-k}$ do not appear in the corresponding term. We use here Convention Eq. (2.11).

Selecting a large enough value for N, as referred to Refs. [24, 130], we expect the solution to the finite hierarchy given by Eq. (9.8) will approximate the solution of the infinite hierarchy of equations

$$\frac{\partial}{\partial t} f^s = \mathcal{B}^s_M\Big(f^s, f^{s+1}, \dots, f^{s+M-1}\Big), \qquad s = 1, 2, \dots, \tag{9.11}$$

where the linear operator $\mathcal{B}^s_M$ is given by

$$\mathcal{B}^s_M\Big(f^s, f^{s+1}, \dots, f^{s+M-1}\Big) = \sum_{m=1}^{M} \mathcal{B}^{s,m}_M f^{s+m-1},$$

and

$$\mathcal{B}_M^{s,m} f^{s+m-1}(v_1, \ldots, v_s) = \sum_{n=1}^{s} \sum_{\{\}} \left(\int_{\mathbb{V}} \right)^{m-1}$$

$$\left(\int_{\mathbb{V}} A^{[m]}(v_n; w, \{v_{s+1}, \ldots v_{s+m-1}\}) \times \right.$$

$$\times \quad a^{[m]}(w, \{v_{s+1}, \ldots, v_{s+m-1}\}) \times$$

$$\times \quad f^{s+m-1}(v_1, \ldots, v_{n-1}, w, v_{n+1}, \ldots, v_s, \ldots, v_{s+m-1})\, dw$$

$$\left. - \quad a^{[m]}(v_n, \{v_{s+1}, \ldots, v_{s+m-1}\})\, f^{s+m-1}(v_1, \ldots, v_{s+m-1}) \right) \times$$

$$\times \quad dv_{s+1} \ldots dv_{s+m-1}\,.$$

We define the admissible hierarchy; see Ref. [130].

Definition 3 *An admissible hierarchy* $\{f^s\}_{s=1,2,3,\ldots}$ *is a sequence of functions* f^s *that satisfy, for* $s = 1, 2, \ldots,$

1. f^s *is a probability density on* $\mathbb{V}^s$,
2. $f^s(v_1, \ldots, v_s) = f^s(v_{r_1}), \ldots, v_{r_s})$ *for all* $v_1, \ldots, v_s$ *in* $\mathbb{V}$ *and for any permutation* $\{r_1, \ldots, r_s\}$ *of the set* $\{1, \ldots, s\}$,
3. $f^s(v_1, \ldots, v_s) = \int_{\mathbb{V}} f^{s+1}(v_1, \ldots, v_{s+1})\, dv_{s+1}$ *for all* $v_1, \ldots v_s$ *in* $\mathbb{V}$.

Then we have; see Ref. [24, 130],

Theorem 20 *Let*

- *Assumptions 10 and 11 be satisfied,*
- $\left\{ \overset{\circ}{f}{}^s \right\}_{s=1,2,\ldots}$ *be an admissible hierarchy.*

Then, for all $t > 0$, *there exists a unique hierarchy* $\{f^s(t)\}_{s=1,2,\ldots}$, $f^s(t) \in L_1^{(s)}$, $s = 1, 2, \ldots,$ *which is a solution of the integral version of Eq.* (9.11) *with initial data* $f^s(0) = \overset{\circ}{f}{}^s$, $s = 1, 2, \ldots.$ *Furthermore,* $\{f^s(t)\}_{s=1,2,\ldots}$, *for all* $t > 0$, *is an admissible hierarchy.*

Proof. The proof can be found in Refs. [24, 130]. ■ QED

The next step is the derivation of the nonlinear equations resulting in the limit $N \to \infty$, that is, the equation at the mesoscopic scale, from Eq. (9.3), cf. [24, 130]). Typically, it is carried out assuming that the initial data are chaotic, that is, factorized probability densities,

$$\overset{\circ}{f}{}^N = \overset{\circ}{f}{}^{N\otimes} := \underbrace{\overset{\circ}{f} \otimes \ldots \otimes \overset{\circ}{f}}_{N\times}, \tag{9.12}$$

where

$$\underbrace{\overset{\circ}{f} \otimes \ldots \otimes \overset{\circ}{f}}_{N\times}(v_1, \ldots, v_N) = \prod_{n=1}^{N} \overset{\circ}{f}(v_n),$$

i.e. N–fold outer product of the probability density $\overset{\circ}{f}$ on V.

As N approaches ∞, the linear equation (9.3) results, as noted in Ref. [24, 130], in a kind of nonlinear Boltzmann–like integro–differential equation that can be related to the mesoscopic scale. As demonstrated in Ref. [130], it can be shown that the chaos property, or factorization property, propagates in time and the solution $f^s(t)$ of the integral version of Eq. (9.11) is the s–product of the solution $f(t)$ of the following nonlinear kinetic equation; cf. Eq. (2.12)

$$\frac{\partial}{\partial t} f(t, v) = \mathcal{G}[f](t, v) - f(t, v)\mathcal{L}[f](t, v), \qquad t > 0, \qquad v \in \mathbb{V}, \tag{9.13}$$

where $\mathcal{G}[f]$ is the gain term, given by

$$\mathcal{G}[f](t, v) = \sum_{m=1}^{M} \left(\int_{\mathbb{V}} \right)^{m-1} A^{[m]}\Big(v; w, \{v_2, \ldots v_m\}\Big) \times$$

$$\times \quad a^{[m]}\Big(w, \{v_2, \ldots, v_m\}\Big) f(t, w) f(t, v_2) \ldots f(t, v_m)\, \mathrm{d}w\, \mathrm{d}v_2 \, \ldots \, \mathrm{d}v_m \,,$$

and $f\,\mathcal{L}[f]$ is the loss term, defined as

$$\mathcal{L}[f](t, v) =$$

$$= \sum_{m=1}^{M} \left(\int_{\mathbb{V}} \right)^{m-1} a^{[m]}\Big(v, \{v_2, \ldots v_m\}\Big) f(t, v_2) \ldots f(t, v_m)\, \mathrm{d}v_2 \ldots \mathrm{d}v_m \,.$$

We adhere to the convention that if $m = 1$ then

$$\sum_{\{\}} \left(\int_{\mathbb{V}} \right)^{m-1}, \qquad \{v_2, \ldots, v_m\}, \qquad f(t, v_2) \ldots f(t, v_m),$$

do not appear in the corresponding term and to Convention Eq. (2.11).

By Theorem 20 we have; see Ref. [130],

Corollary 6 *Let*

- *Assumptions 10 and 11 be satisfied,*
- $\overset{\circ}{f}$ *be a probability density on* $\mathbb{V}$.

Then, for each $\tau > 0$, *there exists an admissible hierarchy* $\{f^s\}_{s=1,2,\ldots}$ *such that*

1. *It is a unique solution of the integral version of Eq.* (9.11) *with chaotic initial data* (9.12),
2. $f^s(t)$ *is chaotic, that is*

$$f^s(t) = \Big(f(t)\Big)^{s\otimes}, \tag{9.14}$$

for all $0 < t \le \tau$ *and* $s = 1, 2, \ldots$, *where* $f(t)$ *is the unique solution in* $L_1^{(1)}$ *of Eq.* (9.13) *with the initial datum* $\overset{\circ}{f}$.

As a by–product, we obtain the existence and uniqueness of solutions to Eq. (9.13).

We are now in a position to present the theorem; referring to Refs. [24, 130], which states the connections between the solutions of Eq. (9.3) and Eq. (9.13), defining a relationship of the microscopic scale with the mesoscopic scale.

Theorem 21 *Let*

- *Assumptions 10 and 11 be satisfied,*
- $\overset{\circ}{f}$ *be a probability density on* $\mathbb{V}$.

Then, for each $\tau > 0$, *there exists a number* N_0 *such that for* $N \geq N_0$

$$\sup_{t\in[0,\tau]} \|f^{N,1}(t) - f(t)\|_1 \leq \frac{\mathfrak{c}}{N^\eta}, \tag{9.15}$$

where $f^N = f^N(t) \in L^1_{(N)}$, $t \in [0, \tau]$, *is the unique nonnegative solution of Eq.* (9.3) *corresponding to the initial datum* (9.12); $f = f(t) \in L^1$, $t \in [0, \tau]$, *is the unique nonnegative solution of Eq.* (9.13) *corresponding to the initial datum* $\overset{\circ}{f}$; *and* η, $\mathfrak{c}$ *are positive constants that depend on* τ.

Theorem 21 asserts that the solution to the nonlinear equation (9.13) on the mesoscopic scale can be approximated by the solutions of the linear equation (9.3) on the microscopic scale, when the value of N is high enough. In contrast, when N is not a large number, both the linear microscopic equation (9.3) and the nonlinear mesoscopic equation (9.13) can independently have a significant impact on the mathematical modelling of phenomena in the life and social sciences, potentially producing different descriptions. We would then obtain two different descriptions, operating in different ranges of applicability, depending on the experimental data. In some cases, we can expect that the microscopic model could provide results that are more closely related to reality. The microscopic description is more detailed and more specific to individual agents.

One can note that complex systems of infinite particles were considered by Y. Kozitsky and collaborators in a series of papers; see Refs. [123–125]. Their stochastic properties and various applications were stated; see also references therein.

9.2 Asymptotic behaviour

In this section, we follow Refs. [131, 132] and we consider the asymptotic in–time behaviour of the solutions to Eq. (9.3). We refer to the notion of partially integral Markov semigroups; cf. Refs. [178, 183, 184, 186], and the lower function theorem of A. Lasota and J. Yorke Ref. [149]; Theorem 2; see also Corollary IV.16 in Ref. [185]—cf. the book by R. Rudnicki and Tyran–Kamińska Ref. [186].

In a fashion analogous to Refs. [132] we can prove the following general result.

Lemma 5 *Let*

- *Assumptions 10 and 11 be satisfied,*
- $a^{[m]}$ *is non–zero for some* $m \in \{1, \ldots, M\}$.

Then the semigroup $\{\exp(t\,\Lambda)\}_{t\geq 0}$ *is a partially integral Markov semigroup.*

Proof. We denote by Γ the following operator

$$\Gamma f = \Lambda f + a^+ f\,,$$

for $f \in L^1_{(N)}$, where $a^+ = \max_{m=1,\ldots,M} a^{[m]}$. We have

$$\Gamma f \geq \Lambda^+ f \geq \max\Big\{0, \Lambda f\Big\} \tag{9.16}$$

and

$$\exp(t\,\Lambda) f = \exp(-a^+ t)\exp(t\,\Gamma) f\,, \tag{9.17}$$

for any probability density f on $\mathbb{V}^N$.

Given a probability density f on $\mathbb{V}^N$, by Eqs. (9.16) and (9.17), it follows

$$\begin{aligned}
&\exp(t\,\Lambda)\, f(v_1, \ldots, v_N) \geq \\
&\geq \tfrac{t^N}{N!}\exp(-a^+ t)\Big(\Lambda^+\Big)^N f(v_1, \ldots, v_N) \geq \\
&\geq \mathfrak{c}_{N,m}(t) \int_{\mathbb{V}^N} b^{[m]}(v_1, \ldots, v_N, w_1, \ldots, w_N) \times \\
&\times f(w_1, \ldots, w_N)\, dw_1 \ldots dw_N\,,
\end{aligned} \tag{9.18}$$

where $\mathfrak{c}_{N,m}(t)$ is a constant that depends on m, n and $t > 0$ and $b^{[m]}$ is a rather complicated function depending on $A^{[m]}$ and $a^{[m]}$,

$$\begin{aligned}
&b^{[m]}(v_1, \ldots, v_N, w_1, \ldots, w_N) = \\
&= A^{[m]}(u_1; v_1, u_2, \ldots u_m) a^{[m]}(w_1, w_2, \ldots, w_m) \times \\
&\times A^{[m]}(v_2; w_2, w_3, \ldots w_{m+1}) a^{[m]}(w_2, u_3, \ldots u_{m+1}) \times \\
&\times \ldots A^{[m]}(v_N; w_N, w_1, \ldots w_{m-1}) a^{[m]}(w_N, w_1, \ldots w_{m-1}).
\end{aligned}$$

Therefore, the semigroup $\{\exp(t\,\Lambda)\}_{t\geq 0}$ is a partially integral Markov semigroup; cf. R. Rudnicki: Ref. [183]. ■ QED

The class of partially integral Markov semigroups, cf. [178, 183, 185], is important in the study of the long–time behaviour of Markov semigroups. However, in order to express the result regarding asymptotic stability, it is necessary to assume stronger conditions.

Theorem 22 *Let*

- *Assumptions 10 and 11 be satisfied;*
- $a^{[m]}$ *be non–zero, for some* $m \in \{1, \ldots, M\}$*;*
- *there exist a measurable nonnegative function* h *on* $\mathbb{V}$ *such that*

$$\int_{\mathbb{V}} h(u)\,\mathrm{d}u > 0\,, \tag{9.19}$$

and

$$A^{[m]}(w, v_1, \ldots, v_m)a^{[m]}(v_1, \ldots, v_m) \geq h(w)\,, \tag{9.20}$$

for any $w, v_1, v_2, \ldots, v_m \in \mathbb{V}$.

Then the stochastic semigroup $\{\exp(t\,\Lambda)\}_{t\geq 0}$ *is asymptotically stable.*

Proof. Given a probability density f on $\mathbb{V}^N$. Analogously as in the proof of Lemma 5 we have

$$\begin{aligned} &\exp\big(t\,\Lambda\big)\,f\big(v_1, \ldots, v_N\big) \;\geq \\ &\geq\; \tfrac{t^N}{N!}\,\exp\big(-a^+ t\big)\,\Big(\Lambda^+\Big)^N f\big(v_1, \ldots, v_N\big) \;\geq \\ &\geq\; \mathfrak{c}_{N,m}(t)\prod_{i=1}^{N} h(u_i)\int_{\mathbb{V}^N} f\big(w_1, \ldots, w_N\big)\,\mathrm{d}w_1 \ldots \mathrm{d}w_N \;= \\ &=\; \mathfrak{c}_{N,m}(t)\prod_{i=1}^{N} h(v_i)\,, \end{aligned} \tag{9.21}$$

where $\mathfrak{c}_{N,m}(t)$ denotes a constant depending on m, n and $t > 0$. Let $\tau > 0$ be fixed, e.g., $\tau = 1$. We have

$$\exp\big(t\,\Lambda\big)\,f \;=\; \exp\big(\tau\,\Lambda\big)\,\exp\big((t-\tau)\,\Lambda\big)\,f\,.$$

Equation (9.21) implies

$$\begin{aligned} &\exp\big(t\,\Lambda\big)\,f\big(v_1, \ldots, v_N\big) \;\geq \\ &\geq\; \mathfrak{c}_{N,m}(\tau)\prod_{i=1}^{N} h(v_i)\Big\|\exp\big((t-\tau)\,\Lambda\big)\,f\Big\|_{L^1_{(N)}}\,, \end{aligned} \tag{9.22}$$

for $t > \tau$. Taking into account that $\exp\big((t-\tau)\,\Lambda\big)f$ is a probability density for each $t > \tau$, we obtain

$$\exp\big(t\,\Lambda\big)\,f\big(v_1, \ldots, v_N\big) \geq \ell(v_1, \ldots, v_N)\,, \tag{9.23}$$

where $\ell(v_1, \ldots, v_N) = \mathfrak{c}_{N,m}(\tau)\prod_{i=1}^{N} h(v_i)$ depends on N but is independent on t and f. By Assumptions 10 and 11, we obtain $\ell \in L^1_{(N)}$. Moreover, Eq. (9.19) yields

$$\int_{\mathbb{V}^N} \ell(v_1, \ldots, v_N)\,\mathrm{d}v_1 \ldots \mathrm{d}v_N > 0\,, \tag{9.24}$$

and ℓ is a *lower function* in the sense of Lasota and Yorke [149]. In fact, the following property holds

$$\lim_{t\to\infty}\Big\|\big(\exp(t\,\Lambda)\,F - \ell\big)^-\Big\|_{L^1_{(N)}} = 0\,,$$

for every probability density F, where $X^- = 0$ if $X \geq 0$ and $X^- = -X$ if $X < 0$. Applying the lower function theorem of Lasota and Yorke, we conclude that the semigroup $\{\exp(t\,\Lambda)\}_{t\geq 0}$ is asymptotically stable. ■ QED

In a specific scenario of the microscopic system that corresponds to the logistic growth on the macroscopic scale, as explained in the next section; cf. also Ref. [132], under specific and somewhat restrictive conditions, it has been shown that any initial probability density, even if it is not initially factorized, eventually converges to an equilibrium probability density that is factorized. This situation is referred to as an asymptotic annihilation of the initial correlations in the system. On the other hand, it was also shown that if the mentioned assumptions are not satisfied, the number of equilibrium states could be large and no annihilation is observed.

9.3 Logistic growth at three scales

In this section, following Refs. [131, 132], we show an example of the mathematical relationships between a simple model on the macroscopic scale and the corresponding models on the mesoscopic and microscopic scales. The starting point—the macroscopic model is probably the simplest, but still significant in many applications, model that is the Verhulst logistic equation, cf. M. Braun Ref. [46]. On a macroscopic scale, it is associated with a limited growth of concentration $\varrho = \varrho(t)$ of agents of some population. The model is a simply nonlinear ordinary differential equation,

$$\frac{\mathrm{d}\,\varrho}{\mathrm{d}\,t} = \alpha\varrho\left(1 - \frac{\varrho}{\kappa}\right), \tag{9.25}$$

where $\alpha > 0$ and $\kappa > 0$ are given parameters. The former is the proliferation rate, and the latter is the carrying capacity that describes the maximum possible concentration under given biological or physical conditions. Given the initial datum

$$0 < \overset{\circ}{\varrho} < \frac{\kappa}{2},$$

the solution of Eq. (9.25) results in a typical S–shaped or logistic curve for $t \geq 0$. These types of curve are typical in a range of disciplines within the Applied Sciences, such as biology, demography, economics, chemistry, and medicine. They can be used, for example, to describe tumour growth. In addition, they appear in geology, social sciences, and neutral networks. For more information, we refer to [27] and the references therein.

It is unlikely that we can develop an alternative straightforward macroscopic model that can accurately fit macroscopic experimental data with such good accuracy regarding the growth of a very simple population exhibiting self–limiting growth, characterized by κ, typically observed in short time intervals.

There are various generalizations in the literature of the simple equation (9.25) into more complex (and interesting) mathematical structures leading to infinite–dimensional dynamical systems including, for example, the space dependence or time delay effects.

The spatial framework can result in Fisher equation, often referred to as the FKPP equation, as well as many of its extensions. The Fisher equation was developed to illustrate the travelling waves of a gene within a population; see Ref. [104]. In addition, more interesting systems incorporate terms related to chemotaxis and haptotaxis, see Refs. [36,152,212] and references therein. The impact of different discrete time delays within the logistic equation was studied in Ref. [43].

According to Refs. [131,132], we study a different mathematical framework that results in an infinite–dimensional system, that is, a microscopic or individual–based model of the type already discussed in this chapter. Keeping in mind the previous part of this chapter, we consider linear evolution equations that correspond to a class of Markov jump processes at the microscopic scale. A microscopic model can still hold considerable significance, even when the number of interacting agents is rather limited. This could be particularly true when the macroscopic model described through the concentration of agents lacks direct interpretation.

As in the previous section, we can show the asymptotic stability for the microscopic model under suitable assumptions; cf. Ref. [132]. When they hold, any initial probability density, regardless of whether it is non–factorized, will eventually converge to an equilibrium probability density that is factorized. However, if the aforementioned assumptions are not satisfied, there are equilibrium probability densities that are not factorized. In other words, initial correlations between individual states may persist.

Similarly to earlier of this chapter, we introduce the set $\mathbb{V}$ of individual states of all agents. For simplicity, as in Refs. [24,131,132], we assume that it is the compact interval $\mathbb{V} = [0, R]$, with some $R > 0$. The parameter $v \in [0, R]$ describes the microscopic individual state of an agent; cf. Section 1.

We define the microscopic model by the generator involving the following functions A and a of the type in Section 9.1; see Ref. [132] and also Refs. [24,131],

Assumption 12 .

- *The rate of interaction*

$$a(v, w) = \frac{\alpha}{\kappa}\, \psi(w)\,, \qquad \forall\ (v, w) \in [0, R]^2\,, \tag{9.26}$$

where ψ is a bounded non–negative measurable function on $[0, R]$;

- *The transition probability*

$$A \in L^1(0, R)\,, \qquad A \geq 0\,, \qquad \int_0^R A(v)\,\mathrm{d}v = 1\,, \tag{9.27}$$

is such that

$$\int_0^R \psi(v)\,A(v)\,\mathrm{d}v = \kappa\,. \tag{9.28}$$

The standard choice of ψ could be

$$\psi(v) = \delta + v\,, \tag{9.29}$$

with some constant $\delta \geq 0$. The case $\delta = 0$ corresponds to the natural situation in which Eq. (9.28) defines the standard first moment. In further considerations, however, see Eq. (9.45) it is necessary to assume $\delta > 0$.

Equations (9.27) and (9.28) determine the relationships between κ, R, A, and ψ. Given $\kappa > 0$ any probability density A defined on $[0, R]$ that satisfies Eq. (9.27) and (9.28) is admissible.

The interpretation of Assumption 12 can be articulated in the following way. The interaction rate a between the test agent is defined by the function ψ depending on the state of another interacting agents. The transition probability A remains unaffected by the states of the agents. Although this assumption represents rather strong simplifications, it may be acceptable in some specific contexts. It would be interesting to generalize this approach to the more general case. However, to some extent, it can probably be considered that the example could probably be the individual domination in the case of bumblebees as considered by E. Jäger and L. Segel Ref. [114].

In continuation of the previous part of this chapter, we take into account a system of finite number $N \in \mathbb{N}$ agents that can interact between them. We define the Markov jump processes through its generator and focus exclusively on the probability densities.

The change in time of probability density is characterized by the following linear equation; cf. Refs. [132],

$$\frac{\partial}{\partial t} f^N = \Lambda_N f^N \,, \tag{9.30}$$

where

$$\Lambda_N f^N = \Lambda_N^+ f^N - \Lambda_N^- f^N \,,$$

and

$$\Lambda_N^+ f^N(t, v_1, v_2, ..., v_N) =$$

$$= \frac{\alpha}{N\kappa} \sum_{\substack{1 \le n,m \le N \\ n \ne m}} A(v_n)\, \psi(v_m) \int_0^R f^N\Big(t, v_1, \ldots, v_{n-1}, w, v_{n+1}, \ldots, v_N\Big)\, dw \,,$$

$$\Lambda_N^- f^N(t, v_1, v_2, ..., v_N) = \frac{\alpha}{N\kappa} \sum_{\substack{1 \le n,m \le N \\ n \ne m}} \psi(v_m)\, f^N(t, v_1, \ldots, v_N) \,.$$

As before, we can show that under Assumption 12 the operator Λ_N is bounded in $L^1_{(N)}$ and generates a stochastic semigroup

$$\Big\{ \exp\big(t\, \Lambda_N\big) \Big\}_{t \ge 0}$$

in $L^1_{(N)}$. Thus, given an initial probability density $F^N \in L^1_{(N)}$, the solution $f^N(t) = \exp\big(t\, \Lambda_N\big) F^N$ exists uniquely and remains a probability density, for every $t > 0$.

We observe that the factorized probability density

$$f^N = \underbrace{A \otimes \ldots \otimes A}_{N\times} \tag{9.31}$$

is an equilibrium solution to Eq. (9.30), i.e. it satisfies

$$\Lambda_N f^N = 0 \,, \qquad f^N \ge 0 \,, \qquad \|f^N\|_{L^1_{(N)}} = 1 \,.$$

The long–time behaviour of solutions to Eq. (9.30) can be analysed in a manner similar to that in the preceding section of this chapter.

At the mesoscopic scale, the corresponding bilinear equation has the form; cf. Ref. [132],

$$\partial_t f(t, v) = \Gamma[f](t, v) \,, \tag{9.32}$$

where

$$\Gamma[f](t, v) = \frac{\alpha}{\kappa} \Big(A(v) f_*(t) f^*(t) - f(t, v) f^*(t) \Big) \,, \tag{9.33}$$

with

$$f_*(t) = \int_0^R f(t,v)\,\mathrm{d}v\,, \tag{9.34}$$

and

$$f^*(t) = \int_0^R \psi(v) f(t,v)\,\mathrm{d}v\,. \tag{9.35}$$

The theory of global existence and uniqueness in L^1 for Eq. (9.32) is standard. The solution $f = f(t)$ represents a probability density; that is,

$$f_*(t) = 1 \qquad \forall\, t > 0\,. \tag{9.36}$$

If the function ψ is defined by Eq. (9.29) with $\delta = 0$, then $\hat{f}$ is the standard first moment of the probability density f.

The solution f to Eq. (13.4.2) is such that

$$\frac{\mathrm{d}}{\mathrm{d}t} f^*(t) = \alpha\, f^*(t) \left(1 - \frac{1}{\kappa}\, f^*(t)\right) \qquad \forall\, t > 0\,, \tag{9.37}$$

that is, $\varrho = f^*$ satisfies the logistic equation (9.25).

Consequently, the characterization on the mesoscopic scale, which is derived from Equation (9.32), simplifies to the macroscopic representation provided by the logistic equation (9.25). Equation (9.32), which generates an infinite–dimensional dynamical system, can be viewed as a straightforward extension of the fundamental 1–dimensional logistic equation. Of course, this is a very special situation related to the form of the equation. In a general situation, significant mathematical difficulties can be expected.

Using Theorem 21, we have; see Ref. [132] and cf. Refs. [24, 131],

Theorem 23 *Let*

- $\alpha > 0,\ \kappa > 0,$
- *Assumption 12 be satisfied,*
- $\overset{\circ}{f}$ *be a probability density on* $[0, R]$.

Then, for any $\tau > 0$*, there exists* N_0 *such that for all* $N > N_0$ *we have*

$$\sup_{t\in[0,\tau]} \left|\, \left(f^{N,1}\right)^* (t) - \varrho(t)\, \right| \leq \frac{\mathfrak{c}}{N^{\varsigma}}\,, \tag{9.38}$$

where $f^N \in L^1_{(N)}$ *is the unique solution of Eq.* (9.30) *corresponding to the factorized initial datum*

$$f^N \Big|_{t=0} = \underbrace{\overset{\circ}{f} \otimes \ldots \otimes \overset{\circ}{f}}_{N\,\times},$$

$f^{N,1} = f^{N,1}(t)$, $t \in [0,\tau]$ *is the "1–agent" marginal probability density, i.e*

$$f^{N,1}(t,v) = \left(\int_0^R\right)^{N-1} f^N(t, v, v_2, \ldots, v_N)\,\mathrm{d}v_2 \cdots \mathrm{d}v_N\,,$$

$\varrho = \varrho(t)$ is the unique solution of Eq. (9.25) *corresponding to the initial datum*

$$\varrho\Big|_{t=0} = \overset{\circ}{f}^{\,*}$$

and ζ, $\mathfrak{c}$ are positive constants that depend on τ.

We have an easy characterization of the time asymptotic behaviour of the solutions to Eq. (9.32); see Theorem 4.2 in Ref. [131].

Theorem 24 *Let*

- *$\alpha > 0$, $\kappa > 0$ be given*
- *Assumption 12 be satisfied,*
- *$\overset{\circ}{f}$ be a probability density on $[0, R]$.*

Then, for all $t > 0$,

$$\|f(t) - A\|_1 \le \| \overset{\circ}{f} - A\|_1 \exp(-\delta\, t)\,, \tag{9.39}$$

where $f(t) \in L^1$ is the unique solution of Eq. (9.32) corresponding to the initial datum

$$f\Big|_{t=0} = \overset{\circ}{f}\,, \tag{9.40}$$

and $\delta = \min\left\{A^*, \left(\overset{\circ}{f}\right)^*\right\} > 0$.

Proof. Following Ref. [131], we observe that $f^*(t)$ satisfies the logistic equation, Eq. (9.25), and therefore

$$\lim_{t\to\infty} f^*(t) = A^* \tag{9.41}$$

We have

- if $\overset{\circ}{f}^{\,*} = A^*$, then $f^*(t) = A^*$ for all $t > 0$,
- if $\overset{\circ}{f}^{\,*} < A^*$, then $f^*(t)$ increases to A^*, $t > 0$,
- if $\overset{\circ}{f}^{\,*} > A^*$, then $f^*(t)$ decreases to A^*, $t > 0$,

Equation (9.32) yields that the solution f in L^1 satisfies

$$\begin{aligned} f(t,v) &= \overset{\circ}{f}(v) \exp\Big(-\int_0^t f^*(s)\,\mathrm{d}s\Big) + \\ &+ A(v) \int_0^t f^*(s) \exp\Big(-\int_s^t f^*(\tilde{s})\,\mathrm{d}\tilde{s}\Big)\,\mathrm{d}s\,, \end{aligned} \tag{9.42}$$

and hence

$$\begin{aligned} f(t,v) &= \overset{\circ}{f}(v) \exp\Big(-\int_0^t f^*(s)\,\mathrm{d}s\Big) + \\ &+ A(v) \left(1 - \exp\Big(-\int_0^t f^*(s)\,\mathrm{d}s\Big)\right). \end{aligned} \tag{9.43}$$

Because

$$\min\left\{A^*\,,\,\overset{\circ}{f}^{\,*}\right\}\le f^*(t)\le \max\left\{A^*\,,\,\overset{\circ}{f}^{\,*}\right\}, \tag{9.44}$$

and $A^* > 0$ as well as $\overset{\circ}{f}^{\,*} > 0$ the proof follows. ■ QED

In order to state the time asymptotic result we apply the lower function theorem like in Theorem 22; see Theorem 3 in Ref. [132].

Theorem 25 *Let*

- *Assumption 12 be satisfied,*
- *There exists a number* $\varepsilon > 0$ *such that*

$$\psi(v)\ge\varepsilon\,,\qquad A(v)\ge\varepsilon\qquad \text{for a.a}\quad v\in[0,R]\,. \tag{9.45}$$

Then the stochastic semigroup $\left\{\exp\left(t\,\Lambda_N\right)\right\}_{t\ge 0}$ *is asymptotically stable.*

From Theorem 25 we obtain the following corollary; see Corollary 1 in Ref. [132],

Corollary 7 *Let*

- *Assumption 12 be satisfied,*
- *Equation* (9.45) *be satisfied,*
- $\overset{\circ}{f}^{\,N}$ *be probability density in* $L^1_{(N)}$.

Then

$$\lim_{t\to\infty}\left\|\exp\left(t\,\Lambda_N\right)\overset{\circ}{f}^{\,N}-\underbrace{A\otimes\ldots\otimes A}_{N\times}\right\|_{L^1_{(N)}}=0\,. \tag{9.46}$$

Corollary 7 shows that any initial probability density $\overset{\circ}{f}^{\,N}$ that is not factorized asymptotically leads to the factorized probability density

$$\underbrace{A\otimes\ldots\otimes A}_{N\times}\,.$$

Equation (9.45) establishes a rather restrictive assumption. However, the subsequent example shows that it is not fundamentally feasible to relax it in the general scenario; see Ref. [144].

Example 4 *We follow Ref. [144] that examines the issue of whether equilibrium solutions related to Eq. (9.3) are unique or not. Thus, we focus on the functions* f^N *such that*

$$f^N\ge 0\,,\qquad \|f^N\|_{L^1_{(N)}}=1\,,\qquad \Lambda_N f^N=0\,.$$

Take, for instance, the simple case $\mathbb{V}=[0,1]$ *and*

$$a\equiv 1\,,$$

$$A(v, w, w_1) = \lambda(v - w), \qquad \forall\, v, w, w_1 \in [0,1], \tag{9.47}$$

where λ *represents a periodic continuous function on the interval* $[0,1]$*, as established by Proposition 3.3 in Ref. [146], and*

$$\lambda \geq 0, \qquad \int_0^1 \lambda(v)\, dv = 1, \qquad (\mathfrak{F}\lambda)_0 = (\mathfrak{F}\lambda)_1 = (\mathfrak{F}\lambda)_{-1} = 1,$$

where $(\mathfrak{F}\lambda)_j$ *is the Fourier coefficient of the function* λ*. Thus, it is evident that all of the functions*

$$f^N(v_1, \ldots, v_N) = 1 + \mathfrak{c}_c \cos\big(2\pi \sum_j^N v_j\big) + \mathfrak{c}_s \sin\big(2\pi \sum_j^N v_j\big), \tag{9.48}$$

if $\mathfrak{c}_c$ *and* $\mathfrak{c}_s$ *are small enough, are equilibrium solutions. Following the construction in Ref. [146] we observe that Assumption* (9.45) *is not satisfied, see Ref. [132]. The asymptotic stability does not hold; see Ref. [144]. Note that none of the functions, defined by Eq. (9.48), is factorized if*

$$|\mathfrak{c}_c| + |\mathfrak{c}_s| \neq 0.$$

Chapter 10

Diauxic growth

10.1 Diauxic

This chapter follows Ref. [133]; see also Ref. [64] and Ph.D. thesis by M. Dębowski [65]. In Section 9.3, it has been noted that logistic growth is observed in experiments in numerous processes of life and social sciences. The corresponding experimental curves are characterized by a sigmoid ("S") shape. This is related to the presence of a unique point of inflection, as is the case of the curve of the normal cumulative distribution function. However, in some situations, more complex behaviour might be experimentally observed, resulting in many points of inflection. Such an observation was emphasized by J. Monod in 1949; see Ref. [159], page 390: "This phenomenon is characterized by a double growth cycle consisting of two exponential phases separated by a phase during which the growth rate passes through a minimum, even becoming negative in some cases".

Jacques Monod used the term "diauxie" to characterize the phenomenon seen in the growth of bacterial cultures. Mathematically, this means the existence of many inflection points for increasing curves. A comparable behaviour was proposed regarding the function of the CDC6 protein in the initiation of mitosis in cells. A new biological hypothesis was proposed on the basis of the experimental data in Ref. [67] and further discussed in Ref. [65]. This hypothesis posits that CDC6 inhibits the activation of inactive complexes of CDK1 and cyclin B during mitotic entry. To formalize this hypothesis, a mathematical model was developed. An additional illustration may involve the DNA melting process, when the base pairs of AT (or TA) and CG (or GC) are organized into two distinct groups, each consisting solely of AT (or TA) and CG (or GC) — cf. Figure 7.14 on page 205 in the book by C. Thompson Ref. [204]. The usual progression of an epidemic, such as, for example, COVID–19, can be represented by curves that exhibit multiple inflection points, referring to the cumulative number of infection cases.

Some mathematical aspects of diauxic growth were studied in Refs. [64, 65].

Definition 4 *An increasing bounded function* $f : [\overset{\circ}{t}, \infty) \to [0, \infty)$ *is referred to as a diauxic growth function, or is said to exhibit diauxic growth, if* f *possesses more than one inflection point for* $t \geq \overset{\circ}{t}$.

10.2 Diauxic at the mesoscopic scale

Although diauxic growth is linked to the macroscopic level, it prompts the question whether models operating at mesoscopic or microscopic scales might also lead to diauxic growth

DOI: 10.1201/9781003537595-10

observable at the macroscopic level. In this section, following Ref. [133], we deal with some preliminary results in this direction. Several nonlinear mesoscopic models are studied which directly result in certain diauxic growth at the macroscopic scale.

In this section, following Ref. [133], we study the behaviour first of the cumulative distribution function and then to the first moment corresponding to the solution of a kinetic (mesoscopic) equation.

For simplicity, we assume that $\mathbb{V} = [0, \infty) \equiv \mathbb{R}_+$ and $\gamma = 3$. Generalizations are straightforward.

In this section, we show that for some assumptions on the parameters A_j, a_j, $j \leq 3$, of Eq. (2.15), the solution $f = f(t, v)$ results in the cumulative distribution function

$$F(t,v) = \int_0^v f(t,\tilde{v})\, \mathrm{d}\tilde{v}\,, \tag{10.1}$$

that has a diauxic growth with respect to $t > 0$, for any sufficiently large $v > 0$. Similarly, we show that the first moment $f^* = f^*(t)$, given by Eq. (9.35), can have a diauxic growth with respect to $t > 0$.

We assume; see Ref. [133],

Assumption 13 *Let*

- a_0, a_1, a_2, a_3 *be such that*

$$\begin{aligned} a_0(v) &= \eta_0\,, \\ a_1(v,w_1) &= \eta_1\, \chi\,(w_1 \leq v)\,, \\ a_2(v,w_1,w_2) &= 2!\, \eta_2\, \chi\,(w_1 \leq v)\, \chi\,(w_2 \leq w_1)\,, \\ a_3(v,w_1,w_2,w_3) &= 3!\, \eta_3\, \chi\,(w_1 \leq v)\, \chi\,(w_2 \leq w_1)\, \chi\,(w_3 \leq w_2)\,, \end{aligned} \tag{10.2}$$

  for all $v, w_1, w_2, w_3 \in \mathbb{R}_+$, *where* χ *is given by Eq. (3.1), and* $\eta_i \geq 0$, $i = 0, 1, 2, 3$, *are constants,*

- A_0 *be such that*

$$\int_0^v A_0\,(\tilde{v}, w)\, \mathrm{d}\tilde{v} = 1\,, \tag{10.3}$$

  for all $v \geq v_0$ *and* $w \in \mathbb{R}_+$, *where* $v_0 > 0$ *is a given constant,*

- A_1, A_2, A_3 *be such that*

$$\begin{aligned} \int_0^v A_1\,(\tilde{v}, w, w_1)\, \mathrm{d}\tilde{v} &= \chi\,(w_1 \leq v)\,, \\ \int_0^v A_2\,(\tilde{v}, w, w_1, w_2)\, \mathrm{d}\tilde{v} &= \chi\,(w_1 \leq v)\,, \\ \int_0^v A_3\,(\tilde{v}, w, w_1, w_2, w_3)\, \mathrm{d}\tilde{v} &= \chi\,(w_1 \leq v)\,, \end{aligned} \tag{10.4}$$

  for all $v, w, w_1, w_2, w_3 \in \mathbb{R}_+$.

We may note that Assumption 13 implies Assumption 5. Simple calculations and Assumption 13 yields

$$\int_0^v \left(\int_0^\infty\right)^j f(t,\tilde{v})\, a_j\,(\tilde{v}, w_1, \dots, w_j) \,\times$$

$$\times \quad f(t,w_1)\dots f(t,w_j)\, \mathrm{d}w_1 \dots \mathrm{d}w_j\, \mathrm{d}\tilde{v} \;= \tag{10.5}$$

$$= \frac{\eta_j}{j+1} \left(\int_0^v f(t,\tilde{v})\, \mathrm{d}\tilde{v}\right)^{j+1},$$

for $j = 1, 2, 3$, any $f(t,\cdot) \in L^1$ and

$$\int_0^v f(t,\tilde{v})\, a_0\,(\tilde{v})\; \mathrm{d}\tilde{v} \;=\; \eta_0 \int_0^v f(t,\tilde{v})\, \mathrm{d}\tilde{v}\,. \tag{10.6}$$

Moreover, for any $f(t,\,\cdot\,) \in L^1$ such that $\|f\|_1 = 1,\ j = 1, 2, 3$, by Assumption 13, we have

$$\int_0^v \left(\int_0^\infty\right)^{j+1} A_j\,(\tilde{v}, w, w_1, \dots, w_j)\, a_j\,(w, w_1, \dots, w_j) \,\times$$

$$\times \quad f(t,w) f(t,w_1)\dots f(t,w_j)\, \mathrm{d}w\, \mathrm{d}w_1 \dots \mathrm{d}w_j\, \mathrm{d}\tilde{v} \;=$$

$$= \; j!\,\eta_j \int_0^\infty f(t,w) \int_0^\infty f(t,w_1) \chi\,(w_1 \le w)\, \chi\,(w_1 \le v) \,\times \tag{10.7}$$

$$\times \quad \int_0^{w_1} f(t,w_2) \dots \int_0^{w_{j-1}} f(t,w_j)\, \mathrm{d}w\, \mathrm{d}w_1 \dots \mathrm{d}w_j\, \mathrm{d}\tilde{v} \;=$$

$$= \; \mathbb{I}_1 + \mathbb{I}_2\,,$$

where

$$\mathbb{I}_1 \;=\; \frac{\eta_j}{j+1} \left(\int_0^v f(t,\tilde{v})\, \mathrm{d}\tilde{v}\right)^{j+1},$$

and

$$\mathbb{I}_2 = \eta_j \left(\left(\int_0^v f(t,\tilde{v})\, \mathrm{d}\tilde{v}\right)^{j} - \left(\int_0^v f(t,\tilde{v})\, \mathrm{d}\tilde{v}\right)^{j+1}\right).$$

Finally, for any $f \in L^1$ such that $\|f\|_1 = 1$, by Assumption 13, for any $v > v_0$, we have

$$\int_0^v \int_0^\infty A_0\,(\tilde{v}, w)\, a_0\,(w)\, f(t,w)\, \mathrm{d}w\, \mathrm{d}\tilde{v} \;=\; \eta_0\,. \tag{10.8}$$

By the above calculations, if we integrate Eq. (2.15) with respect to v, then any solution f of Eq. (2.15), corresponding to an initial datum $\overset{\circ}{f}$, which is a probability density, is such that $\varrho(t) = F(t,v)$ given by Eq. (10.1), for any fixed $v > v_0$, satisfies the following equation

$$\frac{\mathrm{d}}{\mathrm{d}t}\varrho = \mathcal{W}(\varrho)\,, \tag{10.9}$$

where

$$\begin{aligned} \mathcal{W}(x) &= -\eta_3 x^4 + (\eta_3 - \eta_2)\, x^3 + (\eta_2 - \eta_1)\, x^2 + \\ &+ (\eta_1 - \eta_0)\, x + \eta_0 = (1-x)\mathcal{W}_1(x)\,, \end{aligned} \tag{10.10}$$

and

$$\mathcal{W}_1(x) = \eta_3 x^3 + \eta_2 x^2 + \eta_1 x + \eta_0\,.$$

The parameters can be selected so that the function $t \to F(t, v)$ has a diauxic growth for any sufficiently large fixed v. In fact, we have

$$\frac{\mathrm{d}^2}{\mathrm{d}t^2}\varrho = (1-\varrho)\mathcal{W}_1(\varrho)\,\mathcal{W}_2(\varrho)\,,$$

where

$$\mathcal{W}_2(x) = (1-x)\,\frac{\mathrm{d}\,\mathcal{W}_1}{\mathrm{d}\,x}(x) - \mathcal{W}_1(x)\,.$$

We observe that the sign of the second derivative of $\varrho = \varrho(t)$ is defined by the sign of the function $\mathcal{W}_2$. We have

$$\mathcal{W}_2(0) = \eta_1 - \eta_0$$

that is negative under the assumption

$$\eta_1 < \eta_0\,, \tag{10.11}$$

Moreover, we have

$$\mathcal{W}_2(1) = -\,\mathcal{W}_1(1) = -\,(\eta_0 + \eta_1 + \eta_2 + \eta_3) < 0\,, \tag{10.12}$$

and

$$\mathcal{W}_2\left(\frac{1}{2}\right) = \frac{1}{4}\,(\eta_2 + \eta_3) - \eta_0\,. \tag{10.13}$$

Under the assumption

$$\eta_2 + \eta_3 > 4\eta_0 \tag{10.14}$$

we obtain that

$$\mathcal{W}_2\left(\frac{1}{2}\right) > 0\,.$$

Therefore, the solution $\varrho = \varrho(t)$ of Eq. (10.9), which corresponds to the initial datum $\overset{\circ}{\varrho} \in (0,1)$, can only increase. If the conditions that are defined by Eqs. (10.11), (10.14) are satisfied, then there exist points $x_* \in \left(0, \frac{1}{2}\right)$ and $x^* \in \left(\frac{1}{2}\,, 1\right)$ such that

$$\mathcal{W}_2\,(x_*) = \mathcal{W}\,(x^*) = 0\,,$$

and $\mathcal{W}_2$ changes its sign at both points x_* and x_*. The polynomial $\mathcal{W}$ in Eq. (10.9) is such that the unique solution $\varrho = \varrho(t)$ to Eq. (10.9), corresponding to any initial datum $\overset{\circ}{\varrho} \in (0,1)$, satisfies

$$\lim_{t\to\infty} \varrho(t) = 1\,.$$

Moreover, there exist points t_* and t^*, such that

$$\overset{\circ}{t} < t_* < t^*$$

and

$$\varrho(t_*) = x_*, \qquad \varrho(t^*) = x^*.$$

Therefore, we have

Corollary 8 *Let*

- *Assumption 13 be satisfied,*
- $\overset{\circ}{f}$ *be a probability density on* $\mathbb{V} = \mathbb{R}_+$.

Then we can choose the model parameters so that the solution $f = f(t,v)$ *to Eq.* (2.15) *is such that the corresponding* $F = F(t,v)$, *given by Eq.* (10.1), *has the diauxic growth with respect to* t, *for any sufficiently large* $v \in \mathbb{R}_+$.

The above considerations apply to the cumulative distribution function, but a similar result can also be obtained for the first moment. We can use the idea of Section 9.3 and assume that $\mathbb{V} = [0, R]$, for some $R > 0$.

We assume

Assumption 14 *Let*

- a_0, a_1, a_2, a_3 *be such that*

$$\begin{aligned} a_0(v) &= \eta_0, \\ a_1(v, w_1) &= \eta_1\,\psi(w_1), \\ a_2(v, w_1, w_2) &= \eta_2\,\psi(w_1)\,\psi(w_2), \\ a_3(v, w_1, w_2, w_3) &= \eta_3\,\psi(w_1)\,\psi(w_2)\,\psi(w_3), \end{aligned} \tag{10.15}$$

  for all $v, w_1, w_2, w_3 \in \mathbb{R}_+$, *where* η_i *are non–negative constants, and* ψ *is a function as in Assumption 12,*

- A_0, A_1, A_2, A_3 *be such that*

$$\begin{aligned} A_0(v, w) &= A_0(v), \\ A_1(v, w, w_1) &= A_1(v), \\ A_2(v, w, w_1, w_2) &= A_2(v), \\ A_3(v, w, w_1, w_2, w_3) &= A_3(v), \end{aligned} \tag{10.16}$$

  for all $v, w, w_1, w_2, w_3 \in [0, R]$, *and if* $\eta_i > 0$ *then*

$$A_i \geq 0, \qquad \int_0^R A_i(v)\,dv = 1, \qquad \int_0^R \psi(v)\,A_i(v)\,dv = \frac{\tilde{\eta}_i}{\eta_i}, \tag{10.17}$$

  for each $i = 0, 1, 2, 3$, *where* $\tilde{\eta}_i > 0$, $i = 0, 1, 2, 3$, *are constants.*

Similarly as in Section 9.3, $R > 0$ must be in agreement with the parameters of the model, that is, here with $\eta_i, \tilde{\eta}_i$, $i = 0, 1, 2, 3$.

Under Assumption 14, we obtain the macroscopic equation given by Eq. (2.15), with

$$
\begin{aligned}
Q[f](t,v) &= \eta_0 A_0(v) \int_0^R f(t,w)\,\mathrm{d}w - \eta_0 f(t,v) + \\
&+ \eta_1 A_1(v) \left(\int_0^R\right)^2 f(t,w)\psi(w_1)f(t,w_1)\,\mathrm{d}w\,\mathrm{d}w_1 \\
&- \eta_1 f(t,v) \int_0^R \psi(w_1)f(t,w_1)\,\mathrm{d}w + \\
&+ \eta_2 A_2(v) \left(\int_0^R\right)^3 f(t,w)\psi(w_1)f(t,w_1)\psi(w_2)f(t,w_2)\,\mathrm{d}w\,\mathrm{d}w_1\,\mathrm{d}w_2 \\
&- \eta_2 f(t,v) \left(\int_0^R\right)^2 \psi(w_1)f(t,w_1)\psi(w_2)f(t,w_2)\,\mathrm{d}w_1\,\mathrm{d}w_2 + \\
&+ \eta_3 A_3(v) \left(\int_0^R\right)^4 f(t,w)\psi(w_1)f(t,w_1)\psi(w_2)f(t,w_2) \times \\
&\times \psi(w_3)f(t,w_3)\,\mathrm{d}w\,\mathrm{d}w_1\,\mathrm{d}w_2\,\mathrm{d}w_3 \\
&- \eta_3 f(t,v) \left(\int_0^R\right)^3 \psi(w_1)f(t,w_1)\psi(w_2)f(t,w_2) \times \\
&\times \psi(w_3)f(t,w_3)\,\mathrm{d}w_1\,\mathrm{d}w_2\,\mathrm{d}w_3\,,
\end{aligned}
\tag{10.18}
$$

where Convention Eq. (2.11) was used. This equation is a generalization of Eq. (9.32).

Therefore, we take into account the notation given by Eqs. (9.34) and (9.35) and the fact that $f_* \equiv 1$, we obtain the following macroscopic equation, which is satisfied by $\varrho(t) = f^*(t)$, resulting from the mesoscopic equation Eq. (10.18),

$$\frac{\mathrm{d}\,\varrho}{\mathrm{d}\,t} = \mathcal{Y}(\varrho)\,, \tag{10.19}$$

where

$$\mathcal{Y}(x) = -\eta_3 x^4 + (\tilde{\eta}_3 - \eta_2)\,x^3 + (\tilde{\eta}_2 - \eta_1)\,x^2 + (\tilde{\eta}_1 - \eta_0)\,x + \tilde{\eta}_0\,.$$

The polynomial given by Eq. (10.19) is more general than that of Eq. (10.9) and certainly, under suitable assumptions on the parameters, can lead to various diauxic behaviours. Therefore, in such a case, we obtain the diauxic behaviour of the first moment defined by $f^* = f^*(t)$ corresponding to the mesoscopic equation Eq. (2.15) with Q defined by Eq. (10.18).

Therefore, we obtain

Corollary 9 *Let*

- *Assumption 14 be satisfied,*
- $\overset{\circ}{f}$ *be a probability density on* $\mathbb{V} = [0, R]$.

Then we can choose the model parameters so that the solution $f = f(t, v)$ to Eq. (2.15) is such that the corresponding $f^* = f^*(t)$, given by Eq. (9.35), has the diauxic growth with respect to t.

10.3 Diauxic in Game Theory

The following replicator equations are considered in the multi–player games; see, e.g. Ref. [102], together with Refs. [64, 65], and the bibliography therein.

$$\frac{d\varrho}{dt} = \varrho(1-\varrho)\,\mathfrak{P}(\varrho)\,, \tag{10.20}$$

where $\mathfrak{P} = \mathfrak{P}(x)$ is a polynomial. The following two typical examples are examined in Ref. [64]; see also Ref. [65],

$$.\mathfrak{P}(x) = (x-\xi)^2 + \omega\,, \tag{10.21}$$

where $0 < \xi < 1$ and $\omega > 0$ is a small number.

$$\mathfrak{P}(x) = (x-\xi_1)^2(x-\xi_2)^2 + \omega\,, \tag{10.22}$$

where $0 < \xi_1 < \xi_2 < 1$; $(\xi_2 - \xi_1)$ and ω are small numbers. The first Eq. (10.21) pertains to games involving 3 players, while the second Eq. (10.22) pertains to games involving five players. Both are associated with two strategies $\uparrow$ and $\downarrow$ in a population of infinite size. The variables ϱ and $1-\varrho$ represent the frequencies of strategies $\uparrow$ and $\downarrow$, respectively.

For any polynomial $\mathfrak{P} = \mathfrak{P}(x)$ of type

$$\mathfrak{P}(x) = \prod_{i=1}^{m}(x-\xi_i) + \omega\,,$$

where $\omega > 0$, $\xi_i \in (0,1)$, $i = 1, \ldots m$, m is an integer, we can use the elementary inequalities

$$x\,(1-x)\prod_{i=1}^{m}(x-\xi_i) \le x\,(1-x)\,\mathfrak{P}(x) \le x\,(1-x)\prod_{i=1}^{m}(x-\xi_i) + \omega\,,$$
$$\forall\, x \in (0,1)\,,$$

in order to control the number of inflection points of increasing solutions of Eq. (10.20) with appropriately selected parameters ξ_i, $i = 1, \ldots m$ and ω.

Referring to 3 players game the payoff matrix has the form

	$\uparrow\uparrow$	$\uparrow\downarrow$	$\downarrow\downarrow$
$\uparrow$	α_1	α_2	α_3
$\downarrow$	β_1	β_2	β_3 ,

where α_i and β_i, $i = 1, 2, 3$, are the corresponding payoffs. For example, α_2 is the payoff of the first player with strategy $\uparrow$ versus the other two players with strategies $\uparrow$ and $\downarrow$. For simplicity, we can assume that all payoffs are nonnegative.

Let $\rho_\uparrow = \rho_\uparrow(t)$ and $\rho_\downarrow = \rho_\downarrow(t)$ be the densities of players with strategies $\uparrow$ and $\downarrow$, respectively; cf. Ref. [84]. In terms of the average payoffs of the two strategies, their dynamics is defined by the system, Ref. [84],

$$
\begin{aligned}
\tfrac{\mathrm{d}}{\mathrm{d}t}\rho_\uparrow &= \rho_\uparrow\left(\alpha_1\tfrac{\rho_\uparrow^2}{(\rho_\uparrow+\rho_\downarrow)^2} + 2\alpha_2\tfrac{\rho_\uparrow\,\rho_\downarrow}{(\rho_\uparrow+\rho_\downarrow)^2} + \alpha_3\tfrac{\rho_\downarrow^2}{(\rho_\uparrow+\rho_\downarrow)^2} - \kappa\right), \\
\tfrac{\mathrm{d}}{\mathrm{d}t}\rho_\downarrow &= \rho_\downarrow\left(\beta_1\tfrac{\rho_\uparrow^2}{(\rho_\uparrow+\rho_\downarrow)^2} + 2\beta_2\tfrac{\rho_\uparrow\,\rho_\downarrow}{(\rho_\uparrow+\rho_\downarrow)^2} + \beta_3\tfrac{\rho_\downarrow^2}{(\rho_\uparrow+\rho_\downarrow)^2} - \kappa\right),
\end{aligned}
\tag{10.23}
$$

where, in addition to net growth, the linear death terms with the same rate $\kappa > 0$ are taken into account. We see that $\varrho(t) = \frac{\rho_\uparrow(t)}{\rho_\uparrow(t)+\rho_\downarrow(t)}$ satisfies Eq. (10.20) with

$$
\mathfrak{P}(x) = \tilde{\alpha}x^2 + 2\tilde{\beta}x + \zeta\,, \tag{10.24}
$$

where $\tilde{\alpha} = \alpha_3 - 2\alpha_2 + \alpha_1 - \beta_3 + 2\beta_2 - \beta_1$, $\tilde{\beta} = -\alpha_3 + \alpha_2 + \beta_3 - \beta_2$ and $\zeta = \alpha_3 - \beta_3$. Reference [64], see also Ref. [65], states that Eq. (10.21) and Eq. (10.22) can lead to curves with 3 and 5 inflection points, respectively.

Remark 2 *The above methods show that, similarly to Section 9.3, it is possible to formulate microscopic–mesoscopic structures that at the macroscopic level result in the replicator equations in game theory.*

Chapter 11

Discrete states

In numerous life and social science applications, as discussed in Ref. [141], it often makes sense to treat the variables that define the states of population agents as discrete variables. Consequently, the set $\mathbb{V}$ is considered discrete and could be finite or infinite, and the measure μ is the counting measure. For example, when the set $\mathbb{V}$ is finite, we have

$$\mathbb{V} = \Big\{v_1, v_2, \dots, v_m\Big\},$$

where m is an integer, Eq. (2.19) has the form

$$\frac{\mathrm{d}}{\mathrm{d}t} f_j(t) = \zeta\, Q_j[f](t)\,, \qquad j = 1, \dots, m\,, \tag{11.1}$$

$$Q_j[f] = f_j^{\gamma}(t) \sum_{k=1}^{m} a_{kj} f_k(t) - f_j(t) \sum_{k=1}^{m} a_{jk} f_k^{\gamma}(t)\,,$$

and $\zeta = 1$ or $\zeta = -1$.

In particular, for $\zeta = 1$ we obtain

$$\frac{\mathrm{d}}{\mathrm{d}t} f_j(t) = Q_j[f](t)\,, \qquad j = 1, \dots, m\,, \tag{11.2}$$

that is a discrete version of Eq. (2.4). Consequently, in this context, we deal with systems of nonlinear ordinary differential equations. Thus, the mathematical framework is entirely distinct from the integro–differential equations discussed earlier. Nevertheless, the phenomena of organization or disorganization may be covered by these structures—see Ref. [141], where the case of $m = 3$ is studied.

The interpretation of the interaction rates a_{jk}, $j, k = 1, \dots m$, is identical to that in the continuous case: a_{jk} is the coefficient of transition from state j to state k.

As before, we are interested in normalized solutions, that is, we expect that the solutions satisfy

$$\sum_{j=1}^{m} f_j(t) = 1\,, \qquad \forall\, t \geq 0\,. \tag{11.3}$$

We can note that the hyperplane $\sum\limits_{j=1}^{m} f_j = 1$ and the positive cone are invariant.

By Eq. (11.3) we can determine one of the variables, for example f_m, through the remaining variables $f_1, \dots, f_{m-1}$ and consider the problem in the standard (or probability) $m-1$–simplex, similarly like in game theory; see Refs. [58, 108, 109].

The main goal of Ref. [141] is to examine asymmetric interactions between agents, i.e. those in which for a certain pair (j, k), $j, k = 1, \dots, m$, we have

$$a_{jk} \neq a_{k,j}\,.$$

The asymmetric type of interaction is observed in many processes of life and social sciences;

DOI: 10.1201/9781003537595-11

see, e.g., Refs. [2, 20, 110, 161, 197, 202], The asymmetric interactions and asymmetric information in the market is a base of "The Markets for Lemons" theory in economics, as discussed in Ref. [8] by G. Akerlof. When certain agents possess more accurate and detailed information about products on a market than others, this can result in a degradation of the market. In Section 7.2 the case of asymmetric interactions was considered. It has been shown that, in a limit, the kinetic equation approximates a nonlinear inviscid Burgers-type equation, which is often associated with asymmetric scenarios, including models in economics.

The case of discrete particle velocities was considered within the framework of the kinetic theory of gases with possible applications also in other fields. Reference [56] develops a discrete mathematical kinetic theory to model the dynamics of large systems of interacting active particles. A complete reference list up to and including 1985 of the discrete kinetic theory of gases is given in the basic review paper of T. Płatkowski and R. Illner; Ref. [180] and a more recent overview is in the Ph.D. by M.C. Vinerean; Ref. [209].

The situation of 3 possible states, considered in Ref. [141], can be for example related to the Condorcet paradox, or the voting paradox, introduced by the Marquis de Condorcet in the 18th century; see [95, 196, 221], but previously discovered by Ramon Llull in the 13th century. This can manifest itself in the voting process when every voter can support one of the three possible candidates. As a group, preferences form a cyclic pattern, although individual preferences do not have a periodic nature. This typically manifests itself as a ranking of three candidates: the majority of voters prefer A to B, B to C, and C to A; cf. Eq. (11.8). Later, it was developed as Arrow's impossibility theorem; see Refs. [16, 160].

Referring to the mathematical context of Ref. [141], human demographics classes (pyramids) often depict age and gender distributions. Three common types of class include expansive, stationary, and constrictive. An expansive pyramid indicates a higher proportion of young individuals, causing rapid population growth. The stationary class exhibits modest growth, while the constrictive class denotes an older population with declining growth, see Refs. [122, 214] for further details. These characteristics apply to different demographic groups.

In the study of population dynamics, the agents of a population can be arranged in three distinct patterns: they can be evenly spread over an area ("*uniform dispersion*"), scattered unpredictably ("*random dispersion*"), or grouped into clusters ("*clumped dispersion*").

The system 3×3 studied in Ref. [141] lacks periodic solutions that are not constant. In addition, criteria are provided to ensure the asymptotic stability of equilibrium points on the boundary of the simplex. This indicates the asymptotic extinction of two states, which can be interpreted as a type of self-organization. Next, criteria are established for determining the uniqueness and instability of an internal equilibrium point, which means the asymptotic presence of all possible states. The scenario in which $\gamma = 1$ is coupled with asymmetric interaction rates leads to the potentially complex behaviour of solutions of these types of system characterized by asymmetric interactions, which is commonly observed in fields such as economics and certain biological applications, as mentioned above. On the other hand, we note that the system with $\gamma = 1$ and symmetric interaction rates leads to a trivial mathematical structure. Consequently, the intricate dynamics of the system with an asymmetric interaction highlight the significance of differentiating between symmetric and asymmetric interactions.

We can stress that $\gamma = 1$ leads to a variant of the general Lotka–Volterra systems that have been analysed in different contexts, as noted in Refs. [19, 57, 58, 108, 109] and the references therein. In particular, the book by J. Hofbauer and K. Sigmund Ref. [108] is a comprehensive textbook focused on evolutionary games and replicator equations. The replicator equations are equivalent to generalized Lotka–Volterra systems. Reference [58] analyses the replicator equation, offering a classification of stability for three competing strategies based on system parameters. Reference [109] provides a summary of the properties of certain dynamical systems connected to evolutionary game theory. In Ref. [19], S. Baigent examines the qualitative behaviour of several variants of the generalized Lotka–Volterra systems. Specifically, he studies the presence of interior equilibrium states and examines their local stability. Reference [57] provides explicit solutions for heteroclinic trajectories linking nonzero equilibrium points in two–dimensional, noncompetitive Lotka–Volterra systems. The review paper Ref. [172] explores the latest developments in the study of evolutionary dynamics concerning group interactions within structured populations, such as lattices, complex networks, and coevolutionary models. The authors also discuss implications in various fields, including biology, economics, and social sciences, highlighting the integration of statistical physics, network science, and evolutionary game theory.

11.1 Model with 3 states

Following Ref. [141] we can assume that the set $\mathbb{V}$ is composed of 3 elements defining the three possible agent states $\mathbb{V} = \{v_1, v_2, v_3\}$.

Equation (11.2) can be rewritten in the form

$$\begin{aligned}
\frac{\partial}{\partial t} y_1 &= a_{21} y_1^\gamma y_2 + a_{31} y_1^\gamma y_3 - a_{12} y_1 y_2^\gamma - a_{13} y_1 y_3^\gamma \,, \\
\frac{\partial}{\partial t} y_2 &= a_{12} y_2^\gamma y_1 + a_{32} y_2^\gamma y_3 - a_{21} y_2 y_1^\gamma - a_{23} y_2 y_3^\gamma \,, \\
\frac{\partial}{\partial t} y_3 &= a_{13} y_3^\gamma y_1 + a_{23} y_3^\gamma y_2 - a_{31} y_3 y_1^\gamma - a_{32} y_3 y_2^\gamma \,,
\end{aligned} \tag{11.4}$$

where $y_i(t) = f(t, v_i)$, and $a_{ij} = a(v_i, v_j)$, $i, j = 1, 2, 3$.

It is worth observing that the "diagonal" elements vanish. It is intuitive since the transition from a given state to itself is equivalent to staying in the same state; in other words, no real transition appears in such a case.

Assuming that the initial data satisfies

$$\mathring{y}_1 + \mathring{y}_2 + \mathring{y}_3 = 1 \,, \qquad \mathring{y}_i \geq 0 \,, \quad i = 1, 2, 3 \,, \tag{11.5}$$

we observe that the solution should satisfy

$$y_1(t) + y_2(t) + y_3(t) \equiv 1 \,, \qquad y_i(t) \geq 0 \,, \quad i = 1, 2, 3 \,, \qquad \forall\, t \geq 0 \,. \tag{11.6}$$

Equation (11.6) reduces the system of (3×3) ordinary differential equations (ODEs), Eq. (11.4), to the following system of (2×2) ODEs

$$\begin{aligned}
\frac{\partial}{\partial t} y_1 &= a_{21} y_1^\gamma y_2 + a_{31} y_1^\gamma y_3 - a_{12} y_1 y_2^\gamma - a_{13} y_1 y_3^\gamma \,, \\
\frac{\partial}{\partial t} y_2 &= a_{12} y_2^\gamma y_1 + a_{32} y_2^\gamma y_3 - a_{21} y_2 y_1^\gamma - a_{23} y_2 y_3^\gamma \,, \\
y_3 &= 1 - y_1 - y_2 \,,
\end{aligned} \tag{11.7}$$

The situation described by the Condorcet, see the previous part of this chapter, in the framework of the model Eq. (11.7) can be defined by the following conditions

$$a_{12} > a_{21} \,, \quad a_{23} > a_{32} \quad a_{31} > a_{13} \,. \tag{11.8}$$

Reference [141] states

Theorem 26 *Let $\gamma > 1$. Then, within the region $y_1 > 0$, $y_2 > 0$, there is no non-constant periodic solution of Eq. (11.7).*

Proof. The proof follows by the Bendixson–Dulac criterion; see e.g. Theorem 4.8 in Ref. [45]. The Dulac function is

$$B(y_1\,, y_2) = \frac{1}{y_1\, y_2\, (1 - y_1 - y_2)}\,, \qquad y_1\,, y_2 > 0\,:$$

cf. Ref. [141]. ■

Theorem 27 *Consider the equilibrium points*

$$\mathbf{e}_1 = (1, 0)\,, \qquad \mathbf{e}_2 = (0, 1)\,, \qquad \mathbf{e}_3 = (0, 0)$$

of Eq. (11.7). For $i, j, k = 1, 2, 3$, where $i \neq j$, $j \neq k$, and $k \neq i$, if

$$a_{ik} > 0 \qquad \text{and} \qquad a_{jk} > 0\,,$$

then $\mathbf{e}_k$ is (locally) asymptotically stable.

Proof. The proof follows by the direct analysis of the corresponding Jacobi matrices, for details see Ref. [141]. ■

The interpretation of Theorem 27 is immediate. In fact, if the transitions from state i and j, where $i, j = 1, 2, 3$ and $i \neq j$, to state k, where $k = 1, 2, 3$ and $k \neq i$, $k \neq j$, are possible, that is, if $a_{ik} > 0$ and $a_{jk} > 0$, then the equilibrium point $\mathbf{e}_k$ is locally asymptotically stable, which means that the states i and j "locally" vanish in the population.

Detailed calculations of the corresponding Jacobi matrices, in Ref. [141], give the following result.

Theorem 28 *We have*

1. *If $a_{23} > 0$ and $a_{32} > 0$, then there exists a number q_1, $0 < q_1 < 1$, which depends only on a_{23} and a_{32}, such that $\mathbf{q}_1 = (0\,, q_1)$ is an equilibrium point. Moreover, if $a_{12} > 0$ or $a_{13} > 0$, then the equilibrium point $\mathbf{q}_1$ is a saddle point, that is, it is unstable.*

2. *If $a_{13} > 0$ and $a_{31} > 0$, then there exists a number q_2, $0 < q_2 < 1$, which depends only on a_{13} and a_{31}, such that $\mathbf{q}_2 = (q_2\,, 0)$ is an equilibrium point. Moreover, if $a_{21} > 0$ or $a_{23} > 0$, then the equilibrium point $\mathbf{q}_2$ is a saddle point, that is, it is unstable.*

3. *If $a_{12} > 0$ and $a_{21} > 0$, then there exist numbers q_{31} and q_{32}, $q_{3i} > 0$, $i = 1, 2$, $q_{31} + q_{32} = 1$, which depend only on a_{12} and a_{21}, such that $\mathbf{q}_3 = (q_{31}\,, q_{32})$ is an equilibrium point. Moreover, if $a_{31} > 0$ or $a_{32} > 0$, then the equilibrium point $\mathbf{q}_3$ is a saddle point, that is, it is unstable.*

Proof. See the proof of Theorem 2 in Ref. [141]. ■

Theorem 28 states sufficient conditions for the existence of equilibrium points on the boundaries of the simplex

$$\mathbb{S} = \{(y_1, y_2) : \quad y_1 \geq 0\,,\ y_2 \geq 0\,,\ y_1 + y_2 \leq 1\}\,.$$

The point $\mathbf{b}_1$ exists if jumps from state 2 to state 3 and back are possible. It corresponds to the extinction of the state 1 in the population. If, in addition, a jump from 1 to 2 or 3 is possible, then the equilibrium point $\mathbf{b}_1$ is unstable. The interpretation of the remaining equilibrium points is similar.

Laborious calculations in Ref. [141] allow us to determine the conditions for the existence of an internal equilibrium point.

Theorem 29 *Let*

$$a_{21}a_{13}a_{32} = a_{12}a_{31}a_{23}\,. \tag{11.9}$$

Then the existence of a unique equilibrium point $\mathbf{q}_4 = (q_{41}, q_{42})$*, such that*

$$q_{41} > 0\,, \qquad q_{42} > 0\,, \qquad q_{41} + q_{42} < 1\,,$$

is equivalent to the following condition

$$a_{12}a_{21}a_{13}a_{31} > 0 \quad \text{or} \quad a_{12}a_{21}a_{23}a_{32} > 0 \quad \text{or} \quad a_{13}a_{31}a_{23}a_{32} > 0\,. \tag{11.10}$$

The equilibrium point $\mathbf{q}_4$ *is then an unstable node.*

Proof. See the proof of Theorem 3 in Ref. [141]. ■

The condition Eq. (11.9) is weaker than the condition of symmetry of a_{ij}, whereas the condition Eq. (11.10) signifies that at least one of the states must exhibit the possibility of mutual interaction with the remaining states. The latter condition is then essential for the existence of a unique interior equilibrium point.

The discussion of various particular cases related to $\gamma > 1$ can be found in Ref. [141].

If the conditions of Theorem 29 are not satisfied, for example, if

$$a_{12} > 0\,, \qquad \text{but} \qquad a_{13} = a_{31} = a_{23}\,,$$

and

- $a_{21} = 0$, then there are no equilibrium points in the interior of the simplex $\mathbb{S}$,
- $a_{21} > 0$ and $a_{12} = 0$, then there is a segment of equilibrium points in the interior of simplex $\mathbb{S}$.

Proposition 12 *Let Eq.* (11.9) *and* $a_{ij} > 0$*, for all* $i, j = 1, 2, 3,\ i \neq j$ *be satisfied. If* $a_{ij} = a_{kj}$*,* $i, j, k = 1, 2, 3,\ i \neq j,\ i \neq k,\ j \neq k$*, then the separatrix connecting the equilibrium points* $\mathbf{q}_4$ *and* $\mathbf{q}_j$ *is a line segment.*

Proof. See the proof of Proposition 2 in Ref. [141]. ■

The generalization of this result and the characterization of the separatrixes when the conditions of Proposition 12 are not satisfied remains an open problem.

Assuming that $a_{ij} > 0$ for all $i, j = 1, 2, 3$, $i \neq j$ and

$$a_{21} = a_{31}\,, \qquad a_{12} = a_{32}\,, \qquad a_{13} = a_{23}\,,$$

which means that the transition rate from each state to a given state is the same, an explicit formula can be established for the areas of the basins of attraction of the equilibrium points e_1, e_2, and e_3. These areas determine the geometric probabilities of wins for the respective states; see Ref. [141].

Under the assumption that $a_{ij} > 0$ for all $i, j = 1, 2, 3$, $i \neq j$ by Theorems 27 and 28 we conclude that there are three equilibrium points $\mathbf{e}_1$, $\mathbf{e}_2$, and $\mathbf{e}_3$ that are asymptotically stable and three unstable equilibrium points $\mathbf{q}_1$, $\mathbf{q}_2$, and $\mathbf{q}_3$. Moreover, under the same assumptions, there exists an unstable equilibrium point $\mathbf{q}_4$ in the interior of the simplex $\mathbb{S}$.

The case $\gamma = 1$, for asymmetric interaction rates a_{ij}, also leads to interesting behaviour of solutions; cf. Ref. [141]. In Proposition 13 below, without loss of generality, we assume that $a_{12} < a_{21}$ for simplicity.

Standard calculations lead to the following results; see Ref. [141].

Proposition 13 *Let $\gamma = 1$ and $a_{12} < a_{21}$; we have for the system Eq.* (11.7) *the following statements.*

1. *If*

$$a_{13} < a_{31}\,, \qquad a_{23} < a_{32}\,,$$

then the equilibrium point $\mathbf{e}_1$ is asymptotically stable, $\mathbf{e}_2$ is a saddle point and $\mathbf{e}_3$ is an unstable node.

2. *If*

$$a_{13} < a_{31}\,, \qquad a_{32} < a_{23}\,,$$

then the equilibrium point $\mathbf{e}_1$ is asymptotically stable, $\mathbf{e}_2$ is an unstable node and $\mathbf{e}_3$ is a saddle point.

3. *If*

$$a_{31} < a_{13}\,, \qquad a_{32} < a_{23}\,,$$

then the equilibrium point $\mathbf{e}_3$ is asymptotically stable, $\mathbf{e}_1$ is a saddle point and $\mathbf{e}_2$ is an unstable node.

4. *If*

$$a_{31} < a_{13}\,, \qquad a_{23} < a_{32}\,,$$

then the equilibrium points $\mathbf{e}_1$, $\mathbf{e}_2$ and $\mathbf{e}_3$ are saddle points.

Proof. The details of calculations can be found in Section 4 of Ref. [141]. ∎

The condition $a_{ij} > a_{ji}$, for $i, j = 1, 2, 3$, $i \neq j$, means that the possibility of transition from state i to state j is greater than vice versa. The Proposition 13 has a general nature if we exclude some symmetric interactions. Referring to Item 4 of Proposition 13, in this case there exists an equilibrium point $\mathbf{q}_4$ in the interior of the simplex $\mathbb{S}$. This equilibrium point is not hyperbolic, but we can show that the non–constant trajectories inside the simplex are closed curves around the equilibrium point $\mathbf{q}_4$. Then the point $\mathbf{q}_4$ is stable (in the sense of Lyapunov), but it is not asymptotically stable.

The cases allowing some interactions in a symmetrical form, i.e. $a_{ij} = a_{ji}$, for some $i, j = 1, 2, 3$ are discussed in Ref. [141].

Some of the results obtained can be generalized to the multidimensional case $m \geq 4$, but some differences are observed for even and odd m. These issues will be explored in subsequent publications. However, it is worth emphasizing that the discrete case may be as interesting in describing the phenomena of organization and disorganization as the continuous case considered in the previous chapters.

Chapter 12

Space non–local models

12.1 Non–local models with respect to the space variable

The application of both local and non–local in space approaches to the modelling of natural phenomena is a topic that is widely discussed in many scientific fields.

In classical physics, the assumption of a local in–space type of interaction is common and generally accepted. In statistical mechanics, the assumption that two particles collide point–wise, that is, locally, is the basis of Boltzmann theory, leading to the Boltzmann equation. Two particles that are mass points interact, that is, collide, at one point. It is interesting that this simplifying assumption, necessary to create a consistent model, is also the source of basic mathematical difficulties for this equation. In fact, natural *a priori* estimates result from physical properties, the laws of conservation of mass, momentum, and energy, that is, are formulated in the L_1 space setting. But the bilinear collision operator leads outside the L^1 space with respect to the space variable due to the locality of the model; see Ref. [72]. But even in the context of mathematical physics, non–local models were considered; cf. Ref. [13] and references therein.

The non–local nature of quantum mechanics is widely discussed; see e.g. Refs. [147,190] and references therein.

Local models typically result in partial differential equations (PDEs), which are widely accepted mathematical structures in the natural and social sciences. They are commonly accepted as models, but in many cases the spatial structure of individual interacting agents may be important. In the case of invasion of surrounding tissue by cancer, both cancer cells and tissue elements have their own spatial geometrical structure that should be taken into account in proper modelling. Another interesting example of non–local interactions is the creation of flocks in spectacular behaviour of birds; see Refs. [17,21]. In the case of flocks of birds, individual agents interact with each other while maintaining a safe distance between themselves. The distances in such cases seem stochastic in nature. This type of interaction can be modelled by appropriate kernels and the non–local models are formulated in terms of integro–differential equations that contain these kernels. The general references to the non–local models can be found in Refs. [1, 5, 12, 52, 55, 85, 86, 117, 145, 166, 167, 200]. The concept of non–local modelling in biology is discussed in Ref. [150], see also Ref. [94].

It is well known that in various local systems, that is, in terms of partial differential equations, describing cell motions and chemotaxis, the solutions can blow up in a finite time; see, e.g. Refs. [30,36,165,173,174] and references therein. The mathematical structures, integro–differential equations, considered in this monograph can be the basis for the construction of non–local models of various phenomena in life sciences and social sciences. In some cases, such models may have a better mathematical theory of existence and uniqueness, ruling out non–physical blow-ups in those processes in which possible blow–ups in finite time have no justified interpretation.

DOI: 10.1201/9781003537595-12

The non–local, with respect to the spatial variable, models are expressed in the form of integral terms relating to the appropriate non–local interactions and defined by appropriate kernels. These kernels contain parameters that have physical meaning and express the spatial range of interactions. Appropriate transitions to the limiting values, usually to zero values, should lead to local terms in the limit, i.e. a type of localization of description. In such a situation, it could be understood that non–local models could be treated as a generalization of local models; cf. Refs. [94, 150]. If the parameters defining the ranges of interaction are small, non–local models should be "close" to those with local terms. However, "large" values of these parameters could indicate situations far away from those described by local models. As a result, a class of alternative descriptions can be obtained, of which the consistency with the experimental data may be different. Similarly, mathematical properties can be different as well. As a result, we obtain the possibility of an alternative description, whose consistency with different experimental data may be different. Other limit transitions can also be expected to lead to other local equations in terms of partial differential equations that may be relevant in describing various specific situations. Therefore, there are additional opportunities to adapt the models to existing experimental data.

In this monograph, we have already presented some of the space non–local equations. In Section 2.4 we have considered Eq. (2.24) that has a non–local in space structure. Further, in the monograph, the model of the tendon healing process; see Section 13.2; includes non–local space dependence, but does not allow direct movement of agents, that is, collagen fibres, according to the biological situation. Potentially an interesting generalization is the inclusion of direct agent movement; see Section 13.2.3.

12.2 Hybrid models

For the reasons discussed above, an interesting class of equations that can serve as mathematical models of life and social sciences is the class of equations containing non–local terms, i.e. terms expressed through integrals. The literature covering such approaches is enormous. In particular, it includes models that we can call hybrid. These are models that contain both derivatives (local approach) and integral terms (non–local approach).

We will present here only a few approaches of this type, referring the interested reader to the bibliography in the above–mentioned works. We refer here to Refs. [5, 12, 55, 85, 86, 99, 145, 166, 167, 171, 200] and bibliography therein.

Particularly interesting processes, commonly modelled both through local and non–local mathematical structures, are processes related to cell movement including cancer invasion on the surrounding tissue.

Unlike the previously considered equations, the solutions of the equations in this chapter, in accordance with their biological meaning, do not have the property of integral conservation—they do not have to have a probabilistic sense. However, using the strategies of Section 12.4, one can also consider conservative, that is, probabilistic structures.

We assume that $f = f(t, x)$ is the density of cells at time $t \geq 0$ at position $x \in \mathbb{D} \subset \mathbb{R}^{n}$ and $g = g(t, x)$ is the density of tissue at time t at position x. The standard local equation, i.e. in terms of partial differential equations, defining the evolution of function f reads

$$\begin{aligned} \frac{\partial}{\partial t} f(t,x) &= \sigma \frac{\partial}{\partial x} \cdot \Big(f^2(t,x) \frac{\partial}{\partial x} f(t,x) \Big) \\ &- \lambda \frac{\partial}{\partial x} \cdot \Big(f(t,x) \frac{\partial}{\partial x} g(t,x) \Big) + \mathcal{G}_P[f,g](t,x) , \end{aligned} \tag{12.1}$$

σ and λ are positive coefficients. The first, second and third terms on the right–hand side of Eq. (12.1) are a nonlinear diffusion operator, the taxis operator and the third term $\mathcal{G}_P$ is cell proliferation, respectively. The diffusion operator is nonlinear

$$\sigma \frac{\partial}{\partial x} \cdot \left(f^2(t,x) \frac{\partial}{\partial x} f(t,x) \right),$$

with the nonlinearity considered, for example, in Ref. [199]. However, many approaches use linear diffusion or assume other types of nonlinear diffusion. There is a vast literature on various ways of modelling diffusion—see e.g. [63, 207, 217] and references therein, with fundamental objections to linear diffusion formulated, for example, in Ref. [111]. Here, a porous media–like manner is used; cf. Refs. [63, 158, 201, 207]. Typically, proliferation is considered either linearly with respect to function f or logistically, that is, taking into account the limited growth caused by competition.

The equation defining the evolution of function g has the form

$$\frac{\partial}{\partial t} g(t,x) = \mathcal{G}^{[2]}[g,f](t,x), \tag{12.2}$$

where classically the destruction of tissue can be defined by

$$\mathcal{G}^{[2]}[g,f](t,x) \;=\; -\,\eta(x)\, g(t,x)\, f(t,x)\,.$$

or

$$\mathcal{G}^{[2]}[g,f](t,x) \;=\; -g(t,x)\, b \circledast f(t,x)\,,$$

where

$$b \circledast f(x) \;=\; \int_{\mathbb{D}} b(x,y) f(y) \mathrm{d}y$$

is a convolution–like product, η is the rate of degradation and b is the kernel defining degradation.

For various systems describing cell motions and chemotaxis the solutions can blow up in a finite time; see e.g. Chapter 5 in B. Perthame's book Ref. [173], see also [30, 165, 174] and references therein. We can note that the bibliography of Ref. [30] contains 131 items! B. Perthame's book [175] deals with fundamental questions in mathematical biology such as Turing instability, pattern formation, reaction–diffusion systems, and travelling waves.

Reference [200] deals with cancer cell invasion of tissue and focuses on the role of non–local terms that describe competition for space and degradation. The model is a system of reaction–diffusion with taxis partial differential equations with non–local terms related to interactions between cancer cells and host tissue. The model has the form; see Ref. [200],

$$\begin{aligned} \tfrac{\partial}{\partial t} f &= \sigma \Delta u - \lambda \tfrac{\partial}{\partial x} \cdot \left(f \tfrac{\partial}{\partial x} g \right) + \xi_1 f \left(1 - a_{1,1} \circledast f - a_{1,2} \circledast g \right), \\ \tfrac{\partial}{\partial t} g &= -g\, b \circledast f + \xi_2\, g \left(1 - a_{2,1} \circledast f - a_{2,2} \circledast g \right). \end{aligned} \tag{12.3}$$

$f = f(t,x)$ is the cell density and $g = g(t,x)$ is tissue density at time $t \geq 0$ and at position $x \in \mathbb{D}$, as before. The parameters σ and λ are positive constants. The model Eq. (12.3) contains a modified logistic non–local growth term with the constant rate $\xi_1 > 0$. It refers to a neighbourhood of a cell that inhibits proliferation with the kernels $a_{1j} \geq 0$, $j = 1, 2$. Similarly, a modified logistic growth term with constant rates $\xi_2 > 0$ and kernels $a_{2j} \geq 0$, $j = 1, 2$, models tissue production, while b represents the kernel defining tissue degradation. We have

$$\left(a_{ij} \circledast f \right)(x) \;=\; \int a_{ij}(x,y) f(y)\, \mathrm{d}y\,, \quad i, j = 1, 2\,. \tag{12.4}$$

The form of the model is justified by the fact that cancer cells compete for nutrients, e.g. oxygen, with other cells at different spatial locations.

In Ref. [200] bounded domain $\mathbb{D}$ and the no–flux of cancer cells on the boundary of the domain conditions are assumed,

$$\sigma \frac{\partial f}{\partial \nu} - \lambda f \frac{\partial g}{\partial \nu} = 0 , \qquad \text{on} \quad (0, \tau] \times \partial\mathbb{D} ,$$

where ν is the outward normal vector to $\partial\mathbb{D}$.

Using the theory of semigroups, the local existence and uniqueness result in the Sobolev space $W^{l,p}(\mathbb{D})$, where $p > n$, is proved; see Theorem 3.1 in Ref. [200]. Moreover, then it is proved that the solutions to the initial–boundary problem exist globally and are unique in any space dimension n without imposing any kind of smallness conditions on the initial data—see Theorem 4.1 in Ref. [200]. The regularity of the solutions can also be proved and is related to the regularity of the initial data and the kernels (Remark 4.1).

In Ref. [200] the simulations show the influence of non–local terms on the behaviour of solutions. In fact, the shape of the travelling waves of invading cancer cells is related to non–local terms and the size of the parameters that describe the size of non–local interactions. Invasive waves are "classical regular fronts" or present a kind of "overshoot" at the front, indicating a region of high cancer cell density depending on whether the parameters are defined as "close" or "distant", respectively, from the local model.

In the invasion of cancer cells into the surrounding tissue, cancer cells in front of the travelling wave generally do not compete with other cancer cells but only with the cells of the surrounding tissue. This will make them more effective and will cause their concentration to be higher than the assumed limit, in the neighbourhood of the travelling front. This causes the previously mentioned "overshoot" in the front profile and can cause greater effectiveness of the invasion. These numerical results can stress the important role that competition for nutrients (for example, oxygen) and space can play during cancer cell invasion.

Reference [55], following the idea of Ref. [99], studies the situation of three dependent variables: cancer cell density $f = f(t, x)$, tissue density $g = g(t, x)$, and degrading enzyme density $m = m(t, x)$. The model reads

$$\begin{aligned}
\frac{\partial f}{\partial t} &= \underbrace{\sigma_1 \Delta f}_{diffusion} \underbrace{- \frac{\partial}{\partial x} \cdot (f\, a_1 \circledast f)}_{\text{cell-cell adhesion}} \underbrace{- \frac{\partial}{\partial x} \cdot (f\, a_2 \circledast g)}_{\text{cell-tissue adhesion}} + \\
&\quad + \underbrace{\xi_1 f(1 - f - g)}_{\text{proliferation}} , \\
\frac{\partial g}{\partial t} &= - \underbrace{\eta_2 \, g \, m}_{\text{degradation}} , \\
\frac{\partial m}{\partial t} &= \sigma_3 \Delta m + \xi_3 f g - \eta_3 m ,
\end{aligned} \tag{12.5}$$

where σ_1, ξ_1, η_2, σ_3, ξ_3, and η_3 are positive constants that define the coefficients of cancer cell diffusion, cancer cell proliferation, tissue degradation, degrading enzyme diffusion, enzyme proliferation, and enzyme degradation, respectively. The functions a_1, a_2 are $\mathfrak{n}$–dimensional vector–kernels, such that

$$a_i(x, y) = (y - x)\, \tilde{a}_i\,(|y - x|) \,, \qquad i = 1, 2 ,$$

$$\operatorname{supp} \tilde{a}_i \subset [0, \varepsilon] , \qquad \lim_{\varepsilon \to 0} \int_0^{\varepsilon} r^{\mathfrak{n}+1}\, \tilde{a}_i(r)\, \mathrm{d}r = \mathfrak{c} ,$$

where $\mathfrak{c}$ is a constant, $0 < \mathfrak{c} < \infty$, and

$$\lim_{\varepsilon \to 0} \int_0^{\varepsilon} r^{\mathrm{n}+2}\, \tilde{a}_i(r)\, \mathrm{d}r = 0\,.$$

Then we formally have

$$\lim_{\varepsilon \to 0} \left(a_i \circledast f\right)(x) \;=\; \frac{\partial}{\partial x} f(x)\,.$$

In the model Eq. (12.5), some aspects that detail interactions between cancer cells, tissue, and degrading enzymes are taken into account. In particular, we consider

1. cell–cell adhesion,
2. cell–tissue adhesion,
3. tissue degradation by tissue-degrading enzymes,
4. movement of cells and their proliferation,

where cellular adhesion means the binding of a cancer cell to another cancer cell or to a tissue cell.

In Ref. [55] the global–in–time existence–uniqueness result of classical solutions that are uniformly bounded is proved.

The interesting numerical simulation results show that a range of heterogeneous invasive behaviour can be observed depending on the interplay between the cell–cell and cell–tissue parameters.

In Ref. [85] a limit procedure that links non–local models that involve adhesion and chemotaxis to their local versions is proposed. The authors prove a global–in–time existence of weak solutions including the limiting local case. The result does not guarantee the uniqueness of the solutions. The convergence of the non–local terms to the corresponding local terms is also proved.

Reference [199] deals with a model that includes various phenotypes of cancer cells. The transition from the epithelial to mesenchymal (EMT) phenotype and its reverse transition (MET) in the context of cancer progression and spread. The standard cancer invasion model is modified by incorporating the phenotype as a structured variable v, where $v = 0$ corresponds to purely epithelial cancer cells, $v = 1$ corresponds to purely mesenchymal cancer cells, and $0 < v < 1$ represents intermediate phenotypes. Various aspects of the transitions EMT and MET can be found in Refs. [77, 91–93, 101, 121, 179, 203, 220].

Epithelial cells, in the model $v = 0$, are characterized by proliferation and tumour–induced angionesis but do not move. On the other hand, mesenchymal cells, in the model $v = 1$, can move and can be subjected, for example, to diffusion and taxis, as well as product factors, but do not proliferate.

In the model, the following variables are considered $f = f(t,x,v)$, $g = g(t,x)$, and $h = h(t,x)$, which are the densities of cancer cells, tissue, and growth factor TGF–β, respectively.

The complex model that takes into account the existing biological knowledge of the process; see Ref. [199]; is the system of three equations (12.6), (12.7), and (12.8) below

$$
\begin{aligned}
\frac{\partial}{\partial t} f(t,x,v) &= \\
&= \underbrace{H^{\uparrow}(h(t,x)) \left(\int_0^v a(v,w) f(t,x,w)\,\mathrm{d}w - f(t,x,v) \int_v^1 a(v,w)\,\mathrm{d}w \right)}_{\text{EMT transition}} + \\
&+ \underbrace{H^{\downarrow}(h(t,x)) \left(\int_v^1 a(v,w) f(t,x,w)\,\mathrm{d}w - f(t,x,v) \int_0^v a(v,w)\,\mathrm{d}w \right)}_{\text{MET transition}} + \\
&+ \underbrace{\frac{\partial}{\partial x} \cdot \left(\sigma_1(v)\, f^2(t,x,v) \frac{\partial}{\partial x} f(t,x,v) \right)}_{\text{diffusion}} \\
&- \underbrace{\frac{\partial}{\partial x} \cdot \left(\lambda_1(v)\, f(t,x,v) \frac{\partial}{\partial x} g(t,x) \right)}_{\text{haptotaxis}} + \\
&+ \underbrace{\xi_1(v) f(t,x,v) \left(1 - \int_0^1 f(t,x,v)\,\mathrm{d}v - g(t,x) \right)}_{\text{proliferation}} .
\end{aligned} \tag{12.6}
$$

$H^{\uparrow}$ and $H^{\downarrow}$ are regular bounded increasing and decreasing functions of switch type, respectively. The kernel a refers to the nature of changes in the cell's phenotype. The parameters σ_1 and λ_1 are (strictly) positive parameters of diffusion and chemotaxis sensitivity, respectively, that depend on the variable v and we assume that they are regular bounded increasing functions of switch type. An important role here is played by the nonlinear nature of diffusion, assumed in a form of porous media–like manner; cf. Refs. [158,201,207]. The nonlinear non–local diffusion term does not allow for an infinite speed of propagation.

The term defining proliferation is of logistic type with the (strictly) positive parameter ξ_1 of proliferation that depends on the variable v and we assume that it is a regular decreasing function according to the discussion above.

The second equation of the system reads

$$
\begin{aligned}
\frac{\partial}{\partial t} g(t,x) &= \underbrace{-\, g(t,x) \int_0^1 b_2(v) f(t,x,v)\,\mathrm{d}v}_{\text{degradation}} + \\
&+ \underbrace{\xi_2\, g(t,x) \left(1 - \int_0^1 f(t,x,v)\,\mathrm{d}v - g(t,x) \right)}_{\text{ECM remodelling and tissue regrowth}} .
\end{aligned} \tag{12.7}
$$

We assume phenotype–dependent degradation of tissue defined by the kernel b_2 that is a bounded regular positive increasing function of the variable v. Moreover, a logistic type of remodelling and tissue regrowth with a (strictly) positive parameter ξ_2 is assumed.

Finally, to propose the third equation, we assume that the growth factors TGF–β (defined by the density h) are produced by the tissue (with density g) in the presence of cancer cells f, independently of their phenotype, that is, we have the following term

$$\xi_3\, g(t,x) \int_0^1 f(t,x,v)\,\mathrm{d}v\,,$$

where $\xi_3 > 0$ is a constant parameter. The growth factors are small, and the reasonable assumption is that diffusion is linear with the coefficient $\sigma_3 > 0$. We also assume some linear degradation of molecules with the parameter δ_3. Therefore we have

$$\frac{\partial}{\partial t} h(t,x) = \underbrace{\sigma_3 \Delta h(t,x)}_{\text{diffusion}} + \underbrace{\xi_3\, g(t,x) \int_0^1 f(t,x,v)\,\mathrm{d}v}_{\text{production}} - \underbrace{\delta_3 h(t,x)}_{\text{decay}}\,. \tag{12.8}$$

In this way, we obtain the system (12.6), (12.7), (12.8)—the model studied in Ref. [199]. The model contains both integral and differential terms. Movement is described in differential terms, and change in phenotype is described in integral terms. The model was subjected to numerical simulations. They produced interesting results showing the formation of isolated cancer foci in the tissue. It can be assumed that they describe the mechanism of the formation of cancer metastases in places far from the primary tumour. Taking into account cell phenotypes played an important role here.

12.3 Space non–local model of cell movement

In the present section, following Ref. [143], see also Ref. [142], we consider a non–local, with respect to the space variable x, model of cell movement that can be applied, for example, to the invasion of cancer cells into the surrounding tissue. However, the model construction can also be applied in various other situations of specific agent migration processes in life and social sciences.

The non–local way of modelling of biological processes is commonly used; cf. Refs. [145, 167], for example, in the case of cancer invasion on surrounding tissue—see, e.g., [55,85,99, 200] and references therein, as we already discussed in Sections 12.1 and 12.2. It is obvious that some interactions in biology take place at some distances between interacting objects. Even if there are information–transferring factors, taking them into account in models would in most cases radically complicate the models and make them difficult to analyse. Furthermore, obtaining experimental data to validate such modelling would be challenging, if not impossible. For these reasons, the use of non–local models seems justified.

The model, as formulated by Eqs. (12.6), (12.7), (12.8), incorporates different categories of potential agents, each defined by parameter v. We refer to cancer cells and, in particular, to epithelial and mesenchymal cancer cells, cf. [91–93, 100, 101, 121, 179, 203] and references therein, discussed in details in Section 12.2. We want to distinguish two extreme types of cells: epithelial and mesenchymal cells as we already mentioned in Section 12.2. The transition epithelial–mesenchymal is based on a mechanism in which epithelial cells lose their cell polarity and cell–cell adhesion. When they become mesenchymal stem cells, they exhibit an invasive and migratory behaviour. The transition between epithelial and mesenchymal cells (EMT) enables tumour cells to become more invasive. This process is reversible (MET), meaning that the cell can revert back to its original state, and is associated with interactions between tumours and tissues—cf. bibliography in Ref. [199]. This is a process in which cells differentiate into a variety of cell types. The transition is essential for the start of metastasis in cancer progression.

The situation described above is very important for mathematical modelling. It should be emphasised, however, that the strategy proposed here is much more general and can be easily adapted to other agents distinguished by the state defined by the parameter v.

Keeping this in mind and following Refs. [142, 143], as in Section 12.2, we consider the function $f = f(t, x, v)$ that describes the density of cells at the instant of time $t \geq 0$, at position $x \in \mathbb{D}$, where $\mathbb{D}$ for simplicity is assumed to be $\mathfrak{n}$–dimensional torus $\mathbb{T}^{\mathfrak{n}}$. The parameter $v \in [0, 1]$ describes the cell type: for example, as in Section 12.2 in the case of cancer cells, $v = 0$ corresponds to an epithelial–like cancer cell, while $v = 1$ to a mesenchymal–like cancer cell. As we stated previously, the model can be adapted to the situation where the parameter v can specify a different type of moving object including another choice when v is interpreted as the agent velocity.

The basic partial differential equations, with respect to the space variable x, could have the following standard form, cf. Eqs. (12.1) and Eq. (12.6),

$$\begin{aligned} \frac{\partial}{\partial t} f(t,x,v) &= \mathcal{G}_{D,\mathrm{loc}}[f](t,x,v) + \mathcal{G}_{T,\mathrm{loc}}[f,g](t,x,v) + \\ &+ \mathcal{G}_K[f](t,x,v) + \mathcal{G}_P[f,g](t,x,v)\,, \end{aligned} \tag{12.9}$$

where

$$\mathcal{G}_{D,\mathrm{loc}}[f](x,v) = \sigma(v)\,\frac{\partial}{\partial x}\cdot\Big(f^2(x,v)\,\frac{\partial}{\partial x}\,f(x,v)\Big)\,, \tag{12.10}$$

$$\mathcal{G}_{T,\mathrm{loc}}[f,g](x,v) = -\frac{\partial}{\partial x}\cdot\Big(\lambda(x,v)\,f(x,v)\,\frac{\partial}{\partial x}\,g(x)\Big)\,. \tag{12.11}$$

To simplify notation, where ambiguity is not present, we adopt the convention of omitting the explicit indication of the dependence of functions on the variables t and/or v.

The positive coefficients σ and λ of diffusion and taxis can depend on v and (x, v), respectively; however, to simplify, we will assume that these coefficients are constants. Once suitable assumptions are applied, considering variable coefficients does not introduce essential complications.

We can also consider the local operators of diffusion of the form

$$\mathcal{G}_{D,\mathrm{loc},\gamma}[f](x) = \sigma\,\frac{\partial}{\partial x}\cdot\Big(f^{\gamma}(x)\,\frac{\partial}{\partial x}\,f(x)\Big)\,, \tag{12.12}$$

for $\gamma > 0$, and the linear operator

$$\mathcal{G}_{DL,\mathrm{loc}}[f](x) = \sigma\,\Delta_x\,f(x)\,, \tag{12.13}$$

where $\Delta_x = \sum_{j=1}^{\mathfrak{n}} \frac{\partial^2}{\partial x_j^2}$.

The term $\mathcal{G}_K[f] = \mathcal{G}_K^+[f] - f\,\mathcal{G}^-$ has the form

$$\begin{aligned} \mathcal{G}_K^+[f](x,v) &= \int_0^1 a(x,v,w) f(x,w)\,\mathrm{d}w\,, \\ \mathcal{G}_K^-(x,v) &= \int_0^1 a(x,v,w)\,\mathrm{d}w\,. \end{aligned} \tag{12.14}$$

The x–dependence of λ on x reflects the fact of a non–uniform distribution of the orientation of the factors involved in cell movement, for example, tissue in the case of cancer cells; $g(t,x)$ is the density of those factors. The integral terms refer to the possibility of jump from w to v; the term $\mathcal{G}_P$, which may depend on v, describes cell proliferation. Various proliferation terms can be proposed; for example,

- The linear proliferation (independent of g)

$$\mathcal{G}_P[f](x,v) = \xi(v) f(x,v)\,; \tag{12.15}$$

- The *logistic*–like proliferation

$$\mathcal{G}_P[f,g](x,v) = \xi(v) f(x,v)\Big(1 - f(x,v) - g(x)\Big)\,; \tag{12.16}$$

- The non–local proliferation; see Ref. [55],

$$\begin{aligned} \mathcal{G}_P[f,g](x,v) &= \xi(v) f(x,v) \left(1 - \int_{\mathbb{D}} \zeta_1(x,y) f(y,v)\,\mathrm{d}y \right. \\ &- \left. \int_{\mathbb{D}} \zeta_2(x,y) g(y)\,\mathrm{d}y \right). \end{aligned} \tag{12.17}$$

The proliferation term $\mathcal{G}_P$, given by Eq. (12.16) or Eq. (12.17), is nonlinear, and can be related to proliferation–limiting processes in two different ways. The term given by Eq. (12.16) is local with respect to x–variable, whereas the term given by Eq. (12.17) is non–local.

As we mentioned in Section 12.2, in Ref. [199], a more expanded model of cancer invasion was studied to explore how changes in the phenotype of cancer cells can influence the tissue level. The tumour cell phenotype seems to be relevant for disease progression and local invasive spread.

The right–hand side of Eq. (12.9) is composed of 4 terms

1. Nonlinear diffusion term $\mathcal{G}_{D,\mathrm{loc}}[f]$;
2. Nonlinear taxis term $\mathcal{G}_{T,\mathrm{loc}}[f,g]$;
3. Linear term describing the change of cell type $\mathcal{G}_K f$;
4. Proliferation term $\mathcal{G}_P[f,g]$.

The general equation describing the evolution of the factors reads

$$\frac{\partial}{\partial t} g(t,x) = \mathcal{G}^{[2]}[g,f](t,x)\,. \tag{12.18}$$

An equation for the density of surrounding tissue, cf. Eq. (12.7), can be chosen in the following way

$$\mathcal{G}^{[2]}[g,f](x) = -g(x) \int_0^1 b(x,w) f(x,w)\,\mathrm{d}w\,. \tag{12.19}$$

Analogously to Section 12.2, a space–non–local hybrid model can be considered

$$\begin{aligned} \frac{\partial}{\partial t} f(t,x,v) &= \mathcal{G}_{D,\mathrm{loc}}[f](t,x,v) \\ &- \frac{\partial}{\partial x} \cdot \Big(f(t,x,v) \int_{\mathbb{D}} \lambda(x,y,v)\, g(t,y)\,\mathrm{d}y \Big) + \\ &+ \mathcal{G}_K f(t,x,v) + \mathcal{G}_P[f,g](t,x)\,, \\ \frac{\partial}{\partial t} g(t,x) &= \mathcal{G}^{[2]}[g,f](t,x)\,, \end{aligned} \tag{12.20}$$

where $\cdot$ is the scalar product in $\mathbb{R}^{\mathfrak{n}}$ and λ is a vector kernel.

Keeping in mind the structure of Eq. (12.9) and following Ref. [143]; see also Ref. [142], we can propose the corresponding non–local model

$$\begin{aligned} \frac{\partial}{\partial t} f &= \mathcal{G}^{[1]}[f,g]\,, \\ \frac{\partial}{\partial t} g &= \mathcal{G}^{[2]}[g,f]\,, \end{aligned} \tag{12.21}$$

where

$$\mathcal{G}^{[1]}[f,g] = \mathcal{G}_D[f] + \mathcal{G}_T[f,g] + \mathcal{G}_K[f] + \mathcal{G}_P[f,g]\,,$$

and the terms $\mathcal{G}_D[f] = \mathcal{G}_D^+[f] - f^2\,\mathcal{G}_D^- f$ and $\mathcal{G}_T[f,g]$ play the analogous role of diffusion and taxis terms, respectively, in Eq. (12.9), but are expressed in terms of integrals and not partial derivatives. The terms $\mathcal{G}_K f = \mathcal{G}_K^+ f - f\,\mathcal{G}_K^-$ and $\mathcal{G}_P[f,g]$ are the change of type and proliferation terms, respectively, already defined before by Eqs. (12.14) and (12.15) (or Eqs. (12.16) or Eq. (12.17)).

By the results of Section 7.1 we can propose the non–local nonlinear operator acting like diffusion in the form

$$\begin{aligned} \mathcal{G}_D^+[f](x,v) &= f(x,v) \int_{\mathbb{D}} a_D(x,y) f^2(y,v)\mathrm{d}y = \\ &= f(x,v)\big(a_D * f^2\big)(x,v)\,, \\ \mathcal{G}_D^- f(x,v) &= \int_{\mathbb{D}} a_D(x,y) f(y,v)\mathrm{d}y = \big(a * f\big)(x,v)\,, \end{aligned} \tag{12.22}$$

where the support of the kernel a_D is assumed to be concentrated close to the diagonal

$$\Big\{(x,y) \in \mathbb{D}^2 \,:\, x = y\Big\}\,,$$

i.e.,

$$a_D(x,y) = \frac{2\,\sigma}{\varepsilon^{\mathfrak{n}+2}\mathfrak{c}_{\mathfrak{n}}} \chi\big(|x-y| \le \varepsilon\big)\,, \tag{12.23}$$

and $\mathfrak{c}_{\mathfrak{n}}$ is the normalizing constant, i.e.,

$$c_n = \int_{\mathbb{B}(1)} (x_i)^2\,\mathrm{d}x\,, \tag{12.24}$$

i is arbitrary $i = 1, \ldots, \mathfrak{n}$ and $\mathbb{B}(1)$ is the unit ball centred at 0. The parameter $\varepsilon > 0$ is a small parameter that defines the range of interactions.

The results of Section 7.1 show that the term $\mathcal{G}_D[f]$ approximates, as $\varepsilon \to 0$, the first term on the right–hand side of Eq. (12.9), i.e. the diffusion term $\mathcal{G}_{D,\mathrm{loc}}[f]$.

In some situations, instead of a non–local nonlinear diffusion operator $\mathcal{G}_D[f]$ in Eq. (12.22), a non–local linear diffusion operator $\mathcal{G}_{DL}[f]$ can be used corresponding to the standard local linear diffusion operator $\mathcal{G}_{DL,\mathrm{loc}}$ in Eq. (12.13). The operator $\mathcal{G}_{DL}[f]$ in such a case would have the form

$$\mathcal{G}_{DL}[f](x) = \big(a_D * f\big)(x) - f(x)\big(a_D * 1\big)\,, \tag{12.25}$$

where a_D is given by Eq. (12.23).

The local taxis term, i.e. the second term on the right–hand side of Eq. (12.9) $\mathcal{G}_{T,\mathrm{loc}}[f, g]$ satisfies the following a priori properties

1. It is conservative (it does not cause any growth)

$$\int_{\mathbb{D}} \mathcal{G}_{T,\mathrm{loc}}[f, g](t, x)\,\mathrm{d}x = 0\,.$$

2. It does not allow the solution corresponding to non–negative initial data, to be negative.

3. It describes the movement of cells, in terms of $f(t, x)$), towards the highest gradient of g—see [92, 93, 99, 165, 199] and references therein.

It is essential to require that the non–local taxis term meets the above properties. We can suggest it in the form; see Refs. [142, 143],

$$\mathcal{G}_T[f, g] = \mathcal{G}_T^{[1]}[f, g] + \mathcal{G}_T^{[2]}[f, g]\,, \tag{12.26}$$

where—we continue not to indicate the dependence on the variables t and v, for simplicity—

$$\begin{aligned} \mathcal{G}_T^{[1]}[f, g](x) \;&= \\ = \;& \frac{4\,\lambda}{\varepsilon_1\,\varepsilon_2^2\,\varepsilon_3^{\mathfrak{n}-1}} \sum_{j=1}^{\mathfrak{n}} \Bigg(\frac{1}{\varepsilon_1\,\varepsilon_3^{\mathfrak{n}-1}} \int_{\mathbb{D}^2} \chi\Big(0 \le x_j - y_j \le \varepsilon_1\Big)\chi\Big(0 \le z_j - y_j \le \varepsilon_2\Big) \times \\ \times \;& \prod_{i \ne j} \chi\Big(0 \le y_i - x_i \le \varepsilon_3\Big)\chi\Big(0 \le z_i - y_i \le \varepsilon_3\Big)\, f(y)g(z)\,\mathrm{d}z\,\mathrm{d}y \\ - \;& f(x) \int_{\mathbb{D}} \chi\Big(0 \le z_j - x_j \le \varepsilon_2\Big) \prod_{i \ne j} \chi\Big(0 \le z_i - x_i \le \varepsilon_3\Big)\, g(z)\,\mathrm{d}z\,, \end{aligned}$$

and

$$\begin{aligned} \mathcal{G}_T^{[2]}[f, g](x) \;&= \\ = \;& \frac{4\,\lambda}{\varepsilon_2\,\varepsilon_4} \sum_{j=1}^{\mathfrak{n}} \Bigg(\frac{1}{\varepsilon_4\,\varepsilon_3^{\mathfrak{n}-1}} \int_{\mathbb{D}} \chi\Big(0 \le y_j - x_j \le \varepsilon_4\Big) \prod_{i \ne j} \chi\Big(0 \le y_i - x_i \le \varepsilon_3\Big) \times \\ \times \;& f(y)\,g(y)\,\mathrm{d}y - f(x)g(x) \Bigg)\,. \end{aligned}$$

Small ε, ε_1, ε_2, ε_3, ε_4 determine the ranges of interaction. By establishing the suitable relationships between them, one formally arrives at the local taxis term $\mathcal{G}_{T,\mathrm{loc}}[f,g]$.

In the one–dimensional case, that is $\mathfrak{n}=1$, we have

$$\begin{aligned}\mathcal{G}_T^{[1]}[f,g](x) \; &= \\ = \; &\tfrac{4\,\lambda}{\varepsilon_1\,\varepsilon_2^2}\Bigg(\tfrac{1}{\varepsilon_1}\int\limits_{\mathbb{D}}\int\limits_{\mathbb{D}}\chi\Big(0\le x-y\le\varepsilon_1\Big)\chi\Big(0\le z-y\le\varepsilon_2\Big)f(y)g(z)\,\mathrm{d}z\,\mathrm{d}y \\ - \; & f(x)\int\limits_{\mathbb{D}}\chi\Big(0\le z-x\le\varepsilon_2\Big)g(z)\,\mathrm{d}z\Bigg),\end{aligned}$$

and

$$\mathcal{G}_T^{[2]}[f,g](x) = \frac{4\lambda}{\varepsilon_2\,\varepsilon_4}\left(\frac{1}{\varepsilon_4}\int\limits_{\mathbb{D}}\chi(0\le y-x\le\varepsilon_4)f(y)\,g(y)\,\mathrm{d}y - f(x)g(x)\right).$$

In the operators defined above, uniform kernels of interactions have been used to simplify mathematical complications. However, one can also consider more complex kernels defined by small parameters. Then, at the cost of certain mathematical complications, one obtains a more general description in the language of orders of interactions captured by small parameters.

Continuing with the 1–dimensional case, $\mathfrak{n}=1$, we have

$$\begin{aligned}\mathcal{G}_T^{[1]}[f,g](x) \; &= \\ = \; &\tfrac{4\,\lambda}{\varepsilon_1^2\,\varepsilon_2^2}\int\limits_{-\varepsilon_1}^{0}\int\limits_{0}^{\varepsilon_2} f(y+x)\,g(z+y+x)\,\mathrm{d}z\,\mathrm{d}y - \tfrac{4\,\lambda}{\varepsilon_1\,\varepsilon_2^2}\,f(x)\int\limits_{0}^{\varepsilon_2} g(z+x)\,\mathrm{d}z \; = \\ = \; &\tfrac{4\,\lambda}{\varepsilon_1^2\,\varepsilon_2^2}\int\limits_{-\varepsilon_1}^{0}\int\limits_{0}^{\varepsilon_2}\Big(f(x)+yf'(x)+\mathcal{O}(y^2)\Big)\;\times \\ \times \; &\Big(g(x)+(y+z)g'(x)+\tfrac{(y+z)^2}{2}g''(x)+\mathcal{O}((y+z)^3)\Big)\,\mathrm{d}z\,\mathrm{d}y \\ - \; &\tfrac{4}{\varepsilon_1\,\varepsilon_2^2}\,f(x)\int\limits_{0}^{\varepsilon_2}\Big(g(x)+zg'(x)+\tfrac{z^2}{2}g''(x)+\mathcal{O}(z^3)\Big)\,\mathrm{d}z\,.\end{aligned}$$

We observe that

$$\frac{4}{\varepsilon_1^2\,\varepsilon_2^2}\int\limits_{-\varepsilon_1}^{0}\int\limits_{0}^{\varepsilon_2} y(y+z)\,\mathrm{d}z\,\mathrm{d}y = \frac{4\,\varepsilon_1}{3\,\varepsilon_2} - 1\,,$$

$$\frac{4}{\varepsilon_1^2\,\varepsilon_2^2}\int\limits_{-\varepsilon_1}^{0}\int\limits_{0}^{\varepsilon_2}\frac{y^2}{2}\,\mathrm{d}z\,\mathrm{d}y = \frac{2\,\varepsilon_1}{3\,\varepsilon_2}\,,$$

and

$$\frac{4}{\varepsilon_1^2\,\varepsilon_2^2}\int\limits_{-\varepsilon_1}^{0}\int\limits_{0}^{\varepsilon_2}\frac{(y+z)^2}{2}\,\mathrm{d}z\,\mathrm{d}y - \frac{4}{\varepsilon_1\,\varepsilon_2^2}\int\limits_{0}^{\varepsilon_2}\frac{z^2}{2}\,\mathrm{d}z \;=\; \frac{2\,\varepsilon_1}{3\,\varepsilon_2} - 1\,.$$

We aim to establish relationships on the small parameters to achieve

$$\frac{\varepsilon_2\,\varepsilon_1^3}{\varepsilon_1^2\,\varepsilon_2^2} = \frac{\varepsilon_1}{\varepsilon_2} \longrightarrow 0\,, \qquad \frac{\varepsilon_2\,\varepsilon_1^4}{\varepsilon_1^2\,\varepsilon_2^2} = \frac{\varepsilon_1^2}{\varepsilon_2} \longrightarrow 0\,, \qquad \frac{\varepsilon_2^4\,\varepsilon_1}{\varepsilon_1^2\,\varepsilon_2^2} = \frac{\varepsilon_2^2}{\varepsilon_1} \longrightarrow 0\,. \tag{12.27}$$

Moreover, we have to eliminate the following two terms that are present in the expansion of $\mathcal{T}_1[f,g]$

$$f'(x)g(x)\,\frac{4}{\varepsilon_1^2\,\varepsilon_2}\int_{-\varepsilon_1}^{0} y\,\mathrm{d}y = -\frac{2}{\varepsilon_2}\,f'(x)g(x)\,,$$

and

$$f(x)g'(x)\left(\frac{4}{\varepsilon_1^2\,\varepsilon_2^2}\int_{-\varepsilon_1}^{0}\int_{0}^{\varepsilon_2}(y+z)\,\mathrm{d}z\,\mathrm{d}y - \frac{4}{\varepsilon_1\,\varepsilon_2^2}\int_{0}^{\varepsilon_2} y\,\mathrm{d}y\right) = -\frac{2}{\varepsilon_2}\,f(x)g'(x)\,,$$

that is

$$-\frac{2}{\varepsilon_2}\,f'(x)g(x) - \frac{2}{\varepsilon_2}\,f(x)g'(x) = -\frac{2}{\varepsilon_2}\,(fg)'(x)\,. \tag{12.28}$$

The second term $\mathcal{G}_T^{[2]}$ of taxis Eq. (12.26) has only an auxiliary role in adjusting the above–mentioned term given by Eq. (12.28).

As we already mentioned, small parameters ε, ε_1, ε_2, ε_3, ε_4 determine the ranges of interactions. Assuming the suitable relationships between them formally leads to the terms of the local model. Indeed, assuming the following relationships between the small parameters $\varepsilon_1 = \varepsilon^{\frac{3}{2}}$, $\varepsilon_2 = \varepsilon$, $\varepsilon_3 = \varepsilon^2$, and $\varepsilon_4 = \varepsilon^{\frac{3}{2}}$, where $\varepsilon > 0$ is the unique small parameter, we observe a formal convergence of the non–local term of taxis $\mathcal{G}_T[f,g]$ given by Eq. (12.26), as $\varepsilon \to 0$, to the local taxis term $\mathcal{G}_{T,\mathrm{loc}}$ given by Eq. (12.11) with constant λ—see Proposition 14 below. The parameter ε defines the interaction range and shows that the non–local version can be considered as a generalization of the classical local taxis $\mathcal{G}_{T.\mathrm{loc}}$ defined by Eq. (12.11).

On the other hand, we may note that different relationships between parameters may lead to different terms that may be important in describing phenomena in other situations.

Proposition 14 *Let*

- $f \in C^2(\mathbb{D})$ *and* $g \in C^3(\mathbb{D})$,
- $\varepsilon > 0$ *and* $\varepsilon_1 = \varepsilon^{\frac{3}{2}}$, $\varepsilon_2 = \varepsilon$, $\varepsilon_3 = \varepsilon^3$, *and* $\varepsilon_4 = \varepsilon^{\frac{3}{2}}$,

Then

$$\|\mathcal{G}_T[f,g] - \mathcal{G}_{T,\mathrm{loc}}[f,g]\|_{L^\infty(\mathbb{D})} \leq \mathfrak{c}\,\sqrt{\varepsilon}\,, \tag{12.29}$$

where $\mathcal{G}_T$ *is given by Eq.* (12.26), *whereas* $\mathcal{G}_{T,\mathrm{loc}}$ *is given by Eq.* (12.11) *with constant* $\lambda > 0$.

Proof. The proof follows by using Taylor's expansion; see the details in the proof of Theorem 1 in Ref. [143]. ■ QED

We note that the terms $\mathcal{G}_D[f]$, $\mathcal{G}_T[f,g]$, and $\mathcal{G}_K$ are formally conservative and lead to nonnegative solutions for nonnegative initial data. Therefore, $\mathcal{G}_D[f]$ and $\mathcal{G}_T[f,g]$ preserve the essential properties of the corresponding differential term $\mathcal{G}_{D,\mathrm{loc}}[f]$ and $\mathcal{G}_{T,\mathrm{loc}}[f,g]$ given by Eqs. (12.10) and (12.11). The conservative property means that the integrals with respect to $(x,v) \in \mathbb{D} \times [0,1]$ formally vanish. In other words, these terms do not lead to any cell growth or decay according to their physical sense.

As mentioned earlier, other relationships between the small parameters would lead to other terms. In this sense, see Ref. [24]; the non–local model can be considered a more general model than the corresponding local model With the selected parameter relationship, it is reduced to the local model within an appropriate limit. However, if other dependencies result from experimental observations, the local model may not provide satisfactory results. This leads to further mathematical research and references to experiments.

The natural setting for studying the Cauchy Problem for the system Eq. (12.21) is the following Banach space

$$\left(L^1(\mathbb{D}\times[0,1])\cap L^\infty(\mathbb{D}\times[0,1])\right)\times\left(L^1(\mathbb{D})\cap L^\infty(\mathbb{D})\right).$$

In this section, we consider only bounded domains, for example $\mathbb{D}=\mathbb{T}^n$. Consequently, the natural setting for studying the Cauchy problem is the following Banach space $X = L^\infty(\mathbb{D}\times[0,1])\times L^\infty(\mathbb{D})$, equipped with the classical norm

$$\|(f,g)\| = \|f\|_{L^\infty(\mathbb{D}\times[0,1])} + \|g\|_{L^\infty(\mathbb{D})}.$$

In the above considerations, the proposed terms, and in particular their kernels, were such that the limit transition to the local model would be formally realized. However, the existence–uniqueness theory can be developed under more general assumptions. To keep things simple, we do not aim to present the statements in their most general form, in particular the form of the term $\mathcal{G}_T$. Generalizations are, however, straightforward.

We need the following basic assumption.

Assumption 15 *Let*

1. *The kernel a, in Eq.* (12.14), *be a nonnegative bounded measurable function;*
2. *The kernel b, in Eq.* (12.19), *be a nonnegative bounded measurable function;*
3. *The kernels a_D, in Eq.* (12.22), *be a nonnegative bounded measurable functions;*
4. *The proliferation operator $\mathcal{G}_P=\mathcal{G}_P[f,g]$ be locally Lipschitz-continuous operator from X to $L^\infty(\mathbb{D}\times[0,1])$ such that if $(f,g)\in X$ and $f\geq 0$, $g\geq 0$, then*

$$0\leq \mathcal{G}_P[f,g](t,x,v)\leq \mathfrak{c}\, f(t,x,v),$$

with a positive constant $\mathfrak{c}$;

5. *$\lambda\geq 0$, in Eq.* (12.26), *be a constant.*

The following theorem, cf. Theorem 2 in Ref. [143], states the global existence–uniqueness result.

Theorem 30 *Let*

- *$(\overset{\circ}{f},\overset{\circ}{g})\in X$ be such that*

$$\overset{\circ}{f}\geq 0, \qquad \overset{\circ}{g}\geq 0,$$

- *Assumption 15 be satisfied.*

Then, for each $t > 0$, in X there exists a unique solution $(f,g) = (f(t),g(t))$ to the Cauchy problem for Eq. (12.21) *with the initial datum*

$$(f,g)\Big|_{t=0} = (\overset{\circ}{f}, \overset{\circ}{g}).$$

The solution satisfies

$$f(t) \geq 0, \qquad g(t) \geq 0, \tag{12.30}$$

for all $t \geq 0$.

Proof. The operator, defined by the right–hand sides of Eqs. (12.21), is locally Lipschitz–continuous in X. Then there exists $\tau > 0$ such that the local–in–time solution $(f,g) = (f(t),g(t))$, $t \in [0,\tau]$, exists and is unique in X. Moreover, using arguments similar to those in the proof of Theorem 1, see also Theorem 2.2 in Ref. [170], we prove that Eq. (12.30) is satisfied for $t \in [0,\tau]$.

Assume now that $(f,g) = (f(t),g(t))$ is a unique nonnegative solution in X on some interval of time $[0,\tau_1]$, where $\tau_1 \leq \tau$.

By Assumption 15 it follows that

$$\|f(t,.,.)\|_{L^1(\mathbb{D}\times[0,1])} \leq \|\overset{\circ}{f}\|_{L^1(\mathbb{D}\times[0,1])}\, e^{\mathfrak{c}\,t},$$

$t \in [0,\tau_1]$, where $\mathfrak{c}$ is a positive constant.

Equation (12.19) yields

$$g(t,x) \leq \overset{\circ}{g}(x), \qquad t \in [0,\tau_1],\ x \in \mathbb{D}. \tag{12.31}$$

We consider the first term of Eq. (12.21), $\mathcal{G}_D$, multiply by f^{p-1}, where $p > 1$, and integrate with respect to x. We have, cf. Proposition 2,

$$\begin{aligned}
&\int_{\mathbb{D}} f^{p-1}(t,x,v)\,\mathcal{G}_D[f](t,x,v)\,\mathrm{d}x = \\
&= \int_{\mathbb{D}}\int_{\mathbb{D}} a(x,y) f(t,x,v) f(t,y,v)\Big(f^{p-1}(t,x,v) - f^{p-1}(t,y,v)\Big) \times \\
&\times \Big(f(t,y,v) - f(t,x,v)\Big)\,\mathrm{d}y\,\mathrm{d}x \leq 0.
\end{aligned} \tag{12.32}$$

Therefore, the term $\mathcal{G}_D$ does not cause any growth for any $p > 1$.

Consider the second term $\mathcal{G}_T$. Keeping in mind Eq. (12.31) we obtain

$$\begin{aligned}
&\int_{\mathbb{D}} f^{p-1}(t,x,v)\,\mathcal{G}_T[f,g](t,x,v)\,\mathrm{d}x \leq \\
&\leq \mathfrak{c}\,\|f(t,.,v)\|_{L^1(\mathbb{D})} \int_{\mathbb{D}} f^{p-1}(t,x,v)\,\mathrm{d}x \leq \\
&\leq \mathfrak{c}\,\|f(t,.,v)\|^p_{L^p(\mathbb{D})},
\end{aligned} \tag{12.33}$$

for some positive constant $\mathfrak{c}$.

By Young's inequality we obtain

$$\begin{aligned}
&\int_{\mathbb{D}} f^{p-1}(t,x,v)\mathcal{G}_K[f](t,x,v)\,\mathrm{d}x \leq \\
&\leq \mathfrak{c}\,\|f(t,.,v)\|^p_{L^p(\mathbb{D})} + \mathfrak{c}\,\|f(t,.,.)\|^p_{L^p(\mathbb{D}\times[0,1])},
\end{aligned} \tag{12.34}$$

with some constant $\mathfrak{c} > 0$.

Finally the last term $\mathcal{G}_P$ can be estimated by

$$\begin{aligned} \int_{\mathbb{D}} f^{p-1}(t,x,v)\, \mathcal{G}_P[f,g](t,x)\, \mathrm{d}x \;\leq \\ \leq \quad \mathfrak{c}\, \| f(t\,,\,.\,,v) \|^p_{L^p(\mathbb{D})}\,, \end{aligned} \tag{12.35}$$

with some constant $\mathfrak{c} > 0$.

By Assumption 15 and by Eqs. (12.31), (12.32)—(12.35) we obtain

$$\frac{\mathrm{d}}{\mathrm{d}\,t} \| f(t) \|^p_{L^p(\mathbb{D}\times[0,1])} \leq \mathfrak{c}\, \| f(t) \|^p_{L^p(\mathbb{D}\times[0,1])}\,, \tag{12.36}$$

$t \in [0, \tau_1]$ for some constant $\mathfrak{c} > 0$.

Then Grönwall's inequality yields

$$\| f(t) \|^p_{L^p(\mathbb{D}\times[0,1])} \leq \| \overset{\circ}{f} \|^p_{L^p(\mathbb{D}\times[0,1])}\, e^{\mathfrak{c}\,\tau_1}\,, \tag{12.37}$$

for any $t \in [0, \tau_1]$ and any $p \geq 1$, for some constant $\mathfrak{c} > 0$.

Thus

$$\| (f(t), g(t)) \| \leq \| (\, \overset{\circ}{f}\,,\, \overset{\circ}{g}\,) \|\, e^{\mathfrak{c}\,\tau_1}\,, \tag{12.38}$$

for any $t \in [0, \tau_1]$, for some constant $\mathfrak{c} > 0$. This ends the proof. ■ QED

It is interesting that the numerical calculations provided in Figure 4 in Ref. [143] show heterogeneous profiles similar as in a different model with another type of non–localities in Figure 3 in Ref. [55]. This is an interesting inspiration for further research.

Reference [142] in preparation shows that the solutions of the non–local system Eq. 12.21, given by Theorem 30, converge in a weak sense to weak solutions of the local system Eqs. (12.9) and (12.18). This result shows that the non–local model in the appropriate limit reduces to the local model and, in this sense, can be considered more general—potentially corresponding to a larger class of phenomena related to the movement of objects as already mentioned as we already discussed.

12.4 Conservative system for cell movement

The nonlocal system given by Eq. (12.21), like the local system in Eq. (12.9) and the hybrid system in Eq. (12.20), lacks conservation. In the sense that the integral of the right side of the equations does not give 0. This means that the growth caused by the proliferation term in the first equation of the systems and the destruction defined by the second equation, according to the biological meaning, have to influence the behaviour of the solutions. On the other hand, it is worth having conservative mathematical structures, e.g. to give a direct probabilistic interpretation. The idea is known in the theory of stochastic processes; cf., e.g., Example c in Section 8.1 of Ref. [215]. It involves the introduction of an additional artificial subpopulation of the system that has no interpretation, but it is a mental creation that allows the model to be conservative, and therefore probabilistic. Such an idea was used in Ref. [47].

Similarly as in Section 12.3 we consider the function $f = f(t, x, v)$ that describes the density of cells at the instant of time $t \geq 0$, at position $x \in \mathbb{D}$, and with the parameter $v \in [0, 1]$ that describes the cell type; $g(t, x)$ is the density of the factors involved in cell movement, for example, tissue in the case of cancer cells; and $h = h(t)$ is the density of auxiliary sub-population. We can consider the following modification of the system Eq. (12.21)

$$\begin{aligned} \frac{\partial}{\partial t} f &= \mathcal{G}^{[1]}[f, g, h]\,, \\ \frac{\partial}{\partial t} g &= \mathcal{G}^{[2]}[g, f]\,, \\ \frac{\partial}{\partial t} h &= \mathcal{G}^{[3]}[h, f, g]\,, \end{aligned} \tag{12.39}$$

where

$$\mathcal{G}^{[1]}[f, g, h] = \mathcal{G}_D[f] + \mathcal{G}_T[f, g] + \mathcal{G}_K[f] + \mathcal{G}_P[h, f]\,,$$

and $\mathcal{G}_D[f]$, $\mathcal{G}_T[f, g]$, and $\mathcal{G}_K[f]$ are non-local diffusion, non-local taxis, and change of type terms, respectively, and are defined like in Section 12.3; Eqs. (12.22), (12.26), and (12.14). The proliferation term $\mathcal{G}_P[h, f]$ is a modification of Eq. (12.15),

$$\mathcal{G}_P[h, f] = \xi\, h(t)\, f(t, x, v)\,. \tag{12.40}$$

The operator $\mathcal{G}^{[2]}[g, f]$ is given by Eq. (12.19). Finally, the operator $\mathcal{G}^{[3]}[h, f, g]$ is defined by

$$\mathcal{G}^{[3]}[h, f](t) = -\int_{\mathbb{D}} \mathcal{G}^{[2]}[g, f](t, x)\, \mathrm{d}x - \int_{\mathbb{D}} \int_0^1 \mathcal{G}_P[h, f](t, x, v)\, \mathrm{d}v\, \mathrm{d}x\,, \tag{12.41}$$

that is

$$\begin{aligned} \mathcal{G}^{[3]}[h, f](t) &= \int_{\mathbb{D}} \int_0^1 g(x)\, b(x, v) f(t, x, v)\, \mathrm{d}v\, \mathrm{d}x \\ &- \xi\, h(t) \int_{\mathbb{D}} \int_0^1 f(t, x, v)\, \mathrm{d}v\, \mathrm{d}x\,, \end{aligned} \tag{12.42}$$

The system Eq. (12.39) constructed in this way is conservative in nature that is, formally, the solution (f, g, h) is a priori non-negative and we have

$$\begin{aligned} &\int_{\mathbb{D}} \int_0^1 f(t, x, v)\, \mathrm{d}v\, \mathrm{d}x + \int_{\mathbb{D}} g(t, x)\, \mathrm{d}v\, \mathrm{d}x + h(t) = \\ &= \int_{\mathbb{D}} \int_0^1 \overset{\circ}{f}(x, v)\, \mathrm{d}v\, \mathrm{d}x + \int_{\mathbb{D}} \overset{\circ}{g}(x)\, \mathrm{d}v\, \mathrm{d}x + \overset{\circ}{h}\,, \qquad t \geq 0\,, \end{aligned} \tag{12.43}$$

where $(\overset{\circ}{f}, \overset{\circ}{g}, \overset{\circ}{h})$, $\overset{\circ}{f} \geq 0$, $\overset{\circ}{g} \geq 0$, $\overset{\circ}{h} \geq 0$.

Equation (12.43) can be related to the maximum number of all entities in a given system and consequently, after normalization, lead to a probabilistic approach.

In the first two equations of the system Eq. (12.39) the unknown function h appears only in the proliferation term in the first equation. For this reason, using Eq. (12.43), we can express $h = h(t)$ by

$$h(t) = \mathfrak{c} - \left(\int_{\mathbb{D}} \int_0^1 f(t, x, v)\, \mathrm{d}v\, \mathrm{d}x + \int_{\mathbb{D}} g(t, x)\, \mathrm{d}v\, \mathrm{d}x \right), \tag{12.44}$$

where $\mathfrak{c}$ is a given positive constant determined by the initial data. As a result, we will obtain a system of 2 equations for 2 unknown functions

$$\begin{aligned} \frac{\partial}{\partial t} f &= \mathcal{G}_D[f] + \mathcal{G}_T[f,g] + \mathcal{G}_K[f] + \tilde{\mathcal{G}}_P[f,g], \\ \frac{\partial}{\partial t} g &= \mathcal{G}^{[2]}[g,f], \end{aligned} \tag{12.45}$$

where the proliferation term has the form

$$\tilde{\mathcal{G}}_P[f,g] = \xi\, f(t,x,v) \left(\mathfrak{c} - \int_{\mathbb{D}} \int_0^1 f(t,x,v)\, \mathrm{d}v\, \mathrm{d}x - \int_{\mathbb{D}} g(t,x)\, \mathrm{d}v\, \mathrm{d}x \right). \tag{12.46}$$

We can compare Eq. (12.46) with Eq. (12.17).

Let now $\mathcal{G}_D[f]$, $\mathcal{G}_T[f,g]$, $\mathcal{G}_K[f]$ and $\mathcal{G}_P[h,f]$ be non–local diffusion, non–local taxis, change of type and proliferative terms and are defined by Eqs. (12.22), (12.26), (12.14), and (12.40), respectively, The operators $\mathcal{G}^{[2]}[g,f]$ and $\mathcal{G}^{[3]}[h,f,g]$ are defined by Eqs. (12.19) and (12.40). In a similar way to Section 12.3, see Theorem 30, we can state a global existence–uniqueness result.

The Cauchy problem for Eq. (12.39) is considered in the following Banach space $X = L^\infty(\mathbb{D} \times [0,1]) \times L^\infty(\mathbb{D}) \times \mathbb{R}$, equipped with the classical norm

$$\| (f,g,h) \| = \| f \|_{L^\infty(\mathbb{D}\times[0,1])} + \| g \|_{L^\infty(\mathbb{D})} + | h | .$$

Assumption 16 *Let*

1. *The kernel a, in Eq. (12.14), be a nonnegative bounded measurable function;*
2. *The kernel b, in Eq. (12.19), be a nonnegative bounded measurable function;*
3. *The kernels a_D, in Eq. (12.22), be a nonnegative bounded measurable functions;*
4. *$\lambda \geq 0$ in Eq. (12.26) and $\xi \geq 0$ in Eq. (12.40) be constants.*

Then we have the following global–in–time existence–uniqueness result

Theorem 31 *Let*

- $(\overset{\circ}{f}, \overset{\circ}{g}, \overset{\circ}{h}) \in X$ *be such that*

$$\overset{\circ}{f} \geq 0, \qquad \overset{\circ}{g} \geq 0, \qquad \overset{\circ}{h} \geq 0,$$

- *Assumption 16 be satisfied.*

Then, for each $t > 0$, in X there exists a unique solution

$$(f,g,h) = \big(f(t), g(t), h(t)\big),$$

to the Cauchy problem for Eq. (12.39) with the initial datum

$$(f,g,h)\Big|_{t=0} = \left(\overset{\circ}{f}, \overset{\circ}{g}, \overset{\circ}{h} \right).$$

The solution satisfies

$$f(t) \geq 0, \qquad g(t) \geq 0, \qquad h(t) \geq 0, \tag{12.47}$$

and Eq. (12.43), for all $t \geq 0$.

Proof. The proof is analogous to the proof of Theorem 30. ■

Chapter 13

Some applications

In this chapter, we present some interesting models of particular real–world processes. The models are expressed in terms of the integro–differential equations of the type discussed in the monograph. The models presented are preliminary and may be improved as more detailed information about the described processes becomes available. However, they present a general procedure in which processes are described by analysing the interactions between agents of a given population taking into account their interactions. In such processes, integro–differential equations play an important role, providing a chance to capture various aspects of interactions.

The models presented in this chapter are quite diverse in nature. Some are simpler (see, e.g., Section 13.1, and others more complex (see, e.g., Section 13.2) than those considered in the previous part of the monograph. They should be treated as examples of mathematical structures describing certain aspects of the natural processes under consideration. These models should be further developed when new information is obtained from experiments.

In various applications, the microscopic parameter $v \in \mathbb{V}$ can be related to particular individual characteristics of the agents, see Chapter 1. In Ref. [7] it is the general social state of the agent, whereas in Refs. [18] to the pair—the position and velocity of a walker in human crowds. Ref. [34] deals with the dynamics of human crowds, the individual emotional state is taken into account when evacuating a metro station. The framework of kinetic theory in various situations can be found; see, e.g. [7, 31, 32, 34, 130, 139] and references therein. The general references to the mathematical approach of collective phenomena and self–organization in systems composed of large numbers of individuals are Refs. [9, 61, 62, 182]. An interesting application could be the vaccination issues that take into account the psychological individual aspects; see Ref. [213].

One of the most important applications of the models considered in the monograph is the dynamics of opinions, including political opinions, already discussed in Sections 2.3 and 7.2.

As noted in Chapter 1, the various models related to opinion formation were explored in Refs. [126,138,139,169]. References [126,192] deal with opinion dynamics with two types of responses to social influence: conformity and anti–conformity with the aim of illustration of polarized debates in society—see also references therein.

Vehicle traffic can be described taking into account the state of each vehicle and the interactions between vehicles on the individual or microscopic scale. On the other hand, at the level of aggregate flow variables by the density, averaging with respect to velocity at the macroscopic scale; see, for example, [9, 61, 70, 71, 182].

In Ref. [23], see also Ref. [24], the agents are characterized by their position and discrete orientation. Various aspects of swarm behaviour are studied in Refs. [17,21,38,42,48,49,51, 68,115,136,137,155,170].

Generally, there are many possibilities for the applicability of this type of models, in which an individual parameter $v \in \mathbb{V}$ can have many interpretations, in addition to those mentioned above. In biology, for example, it could be the dominance of the individual; see Ref. [114]. In medicine, it could be cell activity; see, e.g., Refs. [24,32,35]. In social sciences;

DOI: 10.1201/9781003537595-13

e.g., social state or socio–economic status; see e.g. Refs. [6, 7, 31, 73, 75, 169], and references therein. It may also be related to position; see Refs. [76, 142, 143] and Chapter 12.1, state of a DNA bond; see Refs. [66, 74]), [fitness and velocity as well.

The study in Ref. [73] deals with the welfare dynamics within societies seen as complex evolutionary systems influenced by various wealth distribution policies, including nonlinear interactions and learning phenomena. Reference [75] is related to emerging collective phenomena within the dynamics of opinions in a society organized into several interacting nodes that form a network. The presented mathematical structure combines two dynamics: the first one on every single node and the second one among the nodes, that is, in the network. The objective is to study how a network structure affects opinion changes. Numerical solutions are used to contrast the emergent behaviours in a consensus–dissent dynamic on an isolated node versus when considering the influence of a network structure connecting interacting nodes.

13.1 Competition between tumour and immune system

As in Ref. [14], we explore a basic kinetic model to illustrate the dynamics of tumour growth in competition with the immune system. This type of modelling might suggest an appropriate therapeutic treatment. The reader is addressed to reviews devoted to modelling, mathematical analysis, numerical simulations, and control of tumour immune system interactions; see, e.g., Refs. [3, 78, 118, 127, 189] and references therein. It is worth mentioning that the literature on the subject is enormous.

The model introduced in [14]; see also [128], is formulated at the cellular level. This model refers to the initial phase of tumour cell emergence and expansion, during which certain endothelial cells lose differentiation and start replicating, interacting with the immune system and surrounding environmental cells. The model is a system of bilinear integro–differential equations. It refers to changes over time of the densities of cell populations related to the activity state of cells. The model includes conservative interactions that change the activity states of cells but preserve the number of cells; proliferative–destructive interactions that change the number of cells; and transition between subpopulations of the system related to the simulation of the immune response of the organism by the production of both active immune cells and macrophages.

The model refers to the following three interacting subpopulations

1. – tumour cells characterized by activity state,
2. – active immune cells,
3. – macrophages,

with densities f_i, $i = 1, 2, 3$, respectively. Furthermore, interactions with the environmental cells with constant density can also be included.

To simplify the model as much as possible, we consider the internal activity state as a relevant factor only for the population of tumour cells. For the other populations, we ignore the internal activity and assume that the functions f_2 and f_3 do not depend on the activity states.

Each tumour cell is characterized by its activation state $v \in [0, 1]$. The function $f_1 = f_1(t, v)$, $f_1 : \mathbb{R}_+ \times [0, 1] \to \mathbb{R}_+$ is the density of tumour cells with activity v at time $t \geq 0$.

The functions $f_i = f_i(t)$, $f_i : \mathbb{R}_+ \to \mathbb{R}_+$, $i = 2, 3$, are the number of active immune cells ($i = 2$) and macrophages ($i = 3$), respectively, at time $t \geq 0$.

Interactions between cells are homogeneous with respect to space and occur instantaneously, that is, without any delay in time. We focus solely on binary interactions that can modify the activation state of cells and affect population size by the destruction or creation of cells.

The basis for the construction of the model is to take into account only the following interactions

- Interactions between tumour and active immune cells that can cause
  - the reduction of the activation state of tumour cells induced by the active immune cells,
  - tumour cells suppressing the activity of active immune cells,
  - the stimulation of the immune response, which involves the generation of new active immune cells;
- Interactions between active tumour cells and macrophages that can cause
  - the elimination of a portion of tumour cells,
  - the suppression of macrophages by tumour cells,
  - the stimulation of the immune response, which involves the generation of macrophages;
- Alterations in tumour cell populations, for example, by interaction with environmental cells,
  - the heightened state of activity of cancer cells,
  - the generation of new tumour cells.

The model is formulated as the following bilinear system of integro–differential, kinetic, equations

$$\begin{aligned} \frac{\partial f_1}{\partial t}(t,v) &= f_2(t)\,\mathcal{A}_{1,2}[f_1](t,v) + A(v)\int_0^1 w f_1(t,w)\,\mathrm{d}w + \\ &+ \mathcal{A}_{1,4}[f_1](t,v) - a_{1,3}\, f_3(t)\, f_1(t,v)\,, \\ \frac{\mathrm{d} f_i}{\mathrm{d} t}(t) &= a^{(i)}\int_0^1 f_1(t,w)\,\mathrm{d}w - a_{i,1}\, f_i(t)\int_0^1 w f_1(t,w)\,\mathrm{d}w\,, \\ i &= 2,3\,, \end{aligned} \tag{13.1}$$

where

- $\mathcal{A}_{1,i}$, $i = 2, 4$, are linear operators and

$$\mathcal{A}_{1,i}[f](v) = \int_0^1 A_{1,i}(v,w)a_{1,i}(w)f(w)\,\mathrm{d}w - a_{1,i}(v)f(v)\,, \tag{13.2}$$

$(v,w) \to A_{1,i}(v,w)a_{1,i}(w)$ and $a_{1,i}$ are nonnegative measurable functions on $[0,1]^2$ and $[0,1]$, respectively,

$$\int_0^1 A_{1,i}(v,w)\,\mathrm{d}v = 1 \qquad \forall\, w \in [0,1]\,, \tag{13.3}$$

$$A_{1,2}(v,w) = 0\,, \qquad \forall\, v,w \in [0,1]\ :\ w \le v\,, \tag{13.4}$$

$$A_{1,4}(v,w) = 0\,, \qquad \forall\, v,w \in [0,1]\ :\ v \le w\,, \tag{13.5}$$

$\mathcal{A}_{1,2}$ characterizes the decrease in tumour activity state resulting from interactions with immune cells, while $\mathcal{A}_{1,4}$ characterizes the increase in tumour activity state. Both terms are conservative; that is, they do not cause any growth of the number of cells, but only a decrease or increase in cell activity.

- A is a nonnegative measurable function that decreases monotonically and $A(1) = 0$, characterizing the proliferative behaviour of tumour cells. The monotonicity of the function is related to the fact that the proliferation rate of new tumour cells with low activation states should be higher than those with high activation states. The corresponding proliferation term, that is, the second term on the right–hand side of the first equation of Eq. (13.1), is proportional to the average activity of tumour cells $\int_0^1 w f_1(t,w)\,\mathrm{d}w$.

- $a_{1,3} \ge 0$ is a constant parameter that characterizes the destruction of tumour cells by macrophages;

- $a^{(i)} \ge 0$, $i = 2$ and $i = 3$, are constant parameters that define the immune response and the rate of macrophage production as a reaction to the presence of tumour cells;

- $a_{i,1} \ge 0$, $i = 2$ and $i = 3$, represent constant parameters that characterize the suppression of immune cell and macrophage production, respectively, by interaction with tumour cells. The corresponding destructive terms, that is, the second terms ($i = 2, 3$) on the right–hand side of the second equation of Eq. (13.1), are, proportional to the average activity of tumour cells $\int_0^1 w f_1(t,w)\,\mathrm{d}w$.

Example 5 *An example of the functions $A_{1,2}$ and $a_{1,2}$ that correspond to the formulated requirements are given in Ref. [14],*

$$A_{1,2}(v,w) = \begin{cases} \frac{2(w-v)}{w^2} & \text{for} \quad v < w\,, \\ 0 & \text{for} \quad v \ge w\,, \end{cases}$$

and

$$a_{1,2}(w) = b_{1,2}w^2\,, \qquad w \in [0,1]\,,$$

where $b_{1,2}$ is a positive constant. Similarly, an example of the functions $A_{1,2}$ and $a_{1,2}$ can be

$$A_{1,4}(v,w) = \begin{cases} \frac{2(v-w)}{(1-w)^2} & \text{for} \quad v > w\,, \\ 0 & \text{for} \quad v \le w\,, \end{cases}$$

and

$$a_{1,4}(w) = b_{1,4}\,(1-w)^2\,, \qquad w \in [0,1]\,,$$

where $b_{1,4}$ *is a positive constant.*

Reference [14] additionally included tumour cell inlet and linear death terms for immune cells and macrophages. For numerical model simulations, the reader is referred to Section 4 of Ref. [14].

We introduce

$$X = \Big\{ (f_1, f_2, f_3) \; : \quad f_1 \in L^1(0,1)\,,\; |f_2| < \infty\,,\; |f_3| < \infty \Big\}$$

the Banach space equipped with the standard norm

$$\|\,(f_1, f_2, f_3)\,\| = \|f_1\|_{L^1(0,1)} + |f_2| + |f_3|\,,$$

and

$$X_+ = \Big\{ (f_1, f_2, f_3) \in X \; : \quad f_1 \geq 0\,,\; f_2 \geq 0\,,\; f_3 \geq 0 \Big\}.$$

Theorem 32 *Let*

- $\left(\overset{\circ}{f}_1\,, \overset{\circ}{f}_2\,, \overset{\circ}{f}_3\right) \in X_+\,,$
- $a_{1,2}\,,\, a_{1,4} \in L^\infty(0,1)\,,$
- $A \in L^1(0,1)\,.$

Then, for every $\tau > 0$*, there exists a unique solution*

$$(f_1\,,\, f_2\,,\, f_3) \in C^0(\,[0,\tau]\,;\, X\,) \cap C^1\,(\,(0,T)\,;\, X\,)$$

of Eq. (13.1) *with the initial datum* $\left(\overset{\circ}{f}_1\,, \overset{\circ}{f}_2\,, \overset{\circ}{f}_3\right)$. *The solution satisfies*

$$(f_1\,,\, f_2\,,\, f_3)\,(t) \in X_+\,, \qquad \forall\, t \in [0,\tau]$$

and

$$\sup_{[0,\tau]} \|\,(f_1\,,\, f_2\,,\, f_3)\,\| \leq \mathfrak{c}_\tau \left\| \left(\overset{\circ}{f}_1\,, \overset{\circ}{f}_2\,, \overset{\circ}{f}_3\right) \right\|$$

for some constant $\mathfrak{c}_\tau$ *depending on* τ.

Proof. See Ref. [14]. By the definition of the system Eq. (13.1) we have

$$\begin{aligned} \tfrac{\mathrm{d}}{\mathrm{d}\,t}\overline{f_1}(t) &\leq \overline{A}\,\overline{f_1}(t)\,, \\ \tfrac{\mathrm{d}}{\mathrm{d}\,t} f_i(t) &\leq a^{(i)}\,\overline{f_1}(t)\,, \qquad i = 2,3\,, \end{aligned} \tag{13.6}$$

where $\overline{f} = \|f\|_{L^1(0,1)}$. ■ QED

Theorem 33 *Let*

- *the conditions of Theorem 32 be satisfied,*
- $(f_1\,,\, f_2\,,\, f_3)$ *be the solution of Eq.* (13.1).

1. *If, for* $i = 2, 3$,

$$a_{i,1} > 0\,, \qquad \overset{\circ}{f}_i \geq \frac{a^{(i)}}{a_{i,1}} \tag{13.7}$$

are satisfied, then

$$f_i(t) \geq \frac{a^{(i)}}{a_{i,1}}\,, \qquad \forall\, t > 0\,.$$

2. *If*

$$a_{3,1} > 0\,, \qquad \overset{\circ}{f}_3 \geq \frac{a^{(3)}}{a_{3,1}}\,, \qquad \bar{A} \leq \frac{a_{1,3}\, a^{(3)}}{b_{3,1}} \tag{13.8}$$

are satisfied, then

$$\bar{f}_1(t) \leq \overline{\overset{\circ}{f}}_1\,, \qquad \forall\, t \geq 0$$

3. *If Eq.* (13.8) *and one of the following two inequalities hold*

$$\overset{\circ}{f}_3 > \frac{a^{(3)}}{a_{3,1}} \qquad \text{or} \qquad \bar{A} < \frac{a_{1,3}\, a^{(3)}}{a_{3,1}}\,,$$

then

$$\lim_{t\to\infty} \bar{f}_1(t) = 0\,.$$

Proof. The proof follows Ref. [14] by the inequalities

$$\bar{f}_1(t) \leq \overline{\overset{\circ}{f}}_1 \exp\left(\bar{A}\, t - a_{1,3} \int_0^t f_3(\tau_1)\, d\tau_1\right), \qquad \forall t \geq 0\,,$$

and

$$f_i(t) \geq \left(\overset{\circ}{f}_i - \frac{a^{(i)}}{a_{i,1}}\right) \exp\left(- a_{i,1} \int_0^t \bar{f}_1(s)\, ds\right) + \frac{a^{(i)}}{a_{i,1}}\,, \qquad \forall\, t > 0\,,$$

for $i = 2, 3$. ■ QED

The above model is very simplified and demands improvement to adequately describe the biological situation appropriately. The literature on models of this type is enormous, both in the context of models of interactions between cancer and the immune system; see, e.g., Refs. [3,78,120], as well as in other applications. Interesting applications of kinetic equations have been used to describe autoimmune diseases. Reference [60] introduced a novel mathematical model to evaluate the effects of drug therapies on autoimmune diseases. An optimal control problem was analysed and numerical tests were performed to explore strategies for optimal treatment. A mathematical model of kinetic type was proposed in Ref. [69, 156] to characterize the typical chronic aspect of autoimmune diseases. This model concerns the cellular interactions of the immune system within the autoimmune context and shows recurring dynamics. The study demonstrates that steady oscillations can emerge through a Hopf bifurcation. Reference [162] deals with the equilibrium states of a nonlinear kinetic system describing the relapsing–remitting stages observed in many chronic autoimmune diseases.

13.2 Tendon recovery process

Tendon rupture represents a significant traumatic event that deserves attention in mathematical modelling; cf. Ref. [83] and references therein. It can lead to a better understanding of the processes involved in tendon healing to effectively suggest optimal therapeutic treatment. A preliminary attempt toward modelling of tendon healing processes was made in Ref. [83]. Following this paper, we present and study a simple model of collagen remodelling that takes place in the last stage of the tendon recovery process. Here we may mean the situation some time after surgical intervention. The model is expressed in terms of an integro–differential equation of the type similar to those discussed in previous chapters of the monograph. The model takes into account tendon structure, i.e. the collagen fibre orientation, which defines a microscopic variable. Reference [83], formulates the global existence versus blow–ups in finite time results depending on initial data, in a way similar to that we have already observed before in the monograph. The blow–ups of the solutions can be interpreted as the alignments in finite time intervals of all collagen fibres, that is, as the recovery without possible scar formation. We can consider this behaviour as another example of self–organization.

In a healthy tendon, the collagen fibres are organized in parallel formations and an injury is based on a disruption of the alignment. The recovery process should involve reconstructing an aligned structure. Yet, in certain instances, such reconstruction might not happen, leading to the formation of scars. An optimal therapeutic strategy might be suggested by improved alignment of collagen fibres. The therapeutic effect improves when they are more aligned. The model proposed in Ref. [83] should be considered an initial step in the development of optimal therapy. The model refers to possible alignments depending on the initial data, but without any external action and taking into account only an internal mechanism of collagen fibres. The results of this mathematical study can offer a recommendation for the best surgical approach to achieve the most favourable initial data configuration depending on medical intervention. This is because the shape of the initial data is determined by a previous medical intervention. The model can be related to a specific stage of recovery after surgery and can provide, regardless of its preliminary nature, some specific indications to the surgeon. Additionally, the model offers significant insights for future investigations.

13.2.1 A simple model

Keeping in mind the theory of previous chapters, we want to define the evolution of probability density $f(t, x, v)$ that describes the distribution of the orientation of collagen, that is, the probability density of finding a collagen fibre in an instant of time $t > 0$ at the point $x \in \mathbb{D}$ and with orientation $v \in \mathbb{V}$, where $\mathbb{D}$, $\mathbb{V}$ are bounded domains in $\mathbb{R}^n$. In contrast to Eq. (2.1), the space structure is taken into account, but the direct movement of collagen fibres is neglected according to the biological context,.

An uncomplicated and possibly unrealistic illustration of the rate of turning may be

$$T_f(y, v; x, w) = f(t, y, v) \,,$$

which leads to the simple nonlinear equation

$$\begin{aligned}\frac{\partial}{\partial t} f(t,x,v) &= \int_{\mathbb{V}} \int_{\mathbb{D}} f(t,y,v)\, f(t,x,w)\, dy\; dw \\ &- f(t,x,v) \int_{\mathbb{V}} \int_{\mathbb{D}} f(t,y,w)\, dy\; dw\,,\end{aligned} \tag{13.9}$$

A more reasonable definition of the turning rate is

$$T_f(y,v;x,w) = a(y,v;x,w) f^{\gamma}(t,y,v)\,,$$

where $\gamma > 1$ describes the impact of the collagen fibre state in a neighbourhood on the collagen fibre state in the given point of space. The impact is actually controlled by tendon cells, which, for simplification purposes, are excluded from the model. The bigger γ, the stronger the impact. With the above definition, we obtain the following general class of equations; cf. Ref. [83];

$$\begin{aligned}\frac{\partial}{\partial t} f(t,x,v) &= \int_{\mathbb{D}} \int_{\mathbb{V}} a(y,v;x,w)\, f^{\gamma}(t,y,v)\, f(t,x,w)\, dy\; dw \\ &- f(t,x,v) \int_{\mathbb{D}} \int_{\mathbb{V}} a(y,w;x,v)\, f^{\gamma}(t,y,w)\, dy\; dw\,.\end{aligned} \tag{13.10}$$

The kernel $a(y,v;x,w)$ is the impact rate of the collagen fibre with orientation w at the point x and another collagen fibre with orientation v located at the point y. The function $a(y,v;x,w)$ has the meaning of the rate of transition from orientation w to orientation v. It seems reasonable to assume that it depends on the distance between the positions $|y-x|$ in such a way that it is concentrated close to the diagonal $y = x$.

It is important to observe that, unlike Eq. (4.1) with $\gamma = 1$, Eqs. (13.9) and (13.10) with $\gamma = 1$ present a non–trivial mathematical structure due to the fact that the variables x and v have a "*mixed*" role within the equation. In other terms, the pair (x, v) should not be considered as a single variable. This equation appears to be an interesting object for mathematical research, regardless of its role in describing the phenomenon of tendon healing. Reference [83] appears to be the first to investigate such a structure, and the equation has not been studied before in the literature. Its nature is somehow different from those considered in the previous chapters.

The simplest choice is $\gamma = 1$, but the case $\gamma \neq 1$ leads to more interesting behaviour involving possibilities of blow–ups in a finite time for $\gamma > 1$—like in Theorems 36 and 37. As we already mentioned, blow–ups can play an interesting role as a mathematical statement of tendon recovery.

Reference [83] states in Theorem 4.1 ($\gamma = 1$), Theorem 4.3 ($\gamma > 1$) and Theorem 4.4 the results that provide a detailed description in the particular case $\gamma = 2$. In addition, numerical calculations play an important role in indicating which shapes of the initial data lead to complete recovery.

Assuming $\gamma = 1$, the right–hand side of Eq. (13.10) defines the following bilinear operator $Q[f]$

$$\begin{aligned}Q[f](t,x,v) &= \int_{\mathbb{D}} \int_{\mathbb{V}} a(y,v;x,w)\, f(t,y,v)\, f(t,x,w)\, dy\; dw \\ &- f(t,x,v) \int_{\mathbb{D}} \int_{\mathbb{V}} a(y,w;x,v)\, f(t,y,w)\, dy\; dw\,,\end{aligned} \tag{13.11}$$

It is easy to see that the operator is Lipschitz—continuous, conservative in $L^1(\mathbb{D} \times \mathbb{V})$ and its structure is such that the non–negativity of the solutions to Eq. (13.10) is preserved under the assumption that $a \in L^\infty\left((\mathbb{D} \times \mathbb{V})^2\right)$ and $a \geq 0$. The following result then holds; cf. Ref. [83].

Theorem 34 *Let*

- $\gamma = 1$;
- $a \in L^\infty\left((\mathbb{D} \times \mathbb{V})^2\right)$ *and* $a \geq 0$;
- $\overset{\circ}{f}$ *be a probability density in* $\mathbb{D} \times \mathbb{V}$.

Then, for any $t > 0$*, there exists a unique function* $f = f(t)$ *in* $L^1(\mathbb{D} \times \mathbb{V})$ *that is a solution to Eq. (13.10), corresponding to the initial datum* $\overset{\circ}{f}$*. Furthermore,* $f = f(t)$ *is a probability density for each* $t > 0$.

We consider the following marginal probability densities.

$$\varrho(t, x) = \int_{\mathbb{V}} f(t, x, w)\, \mathrm{d}w\,,$$

and

$$\varsigma(t, v) = \int_{\mathbb{D}} f(t, y, v)\, \mathrm{d}y\,.$$

The solutions of Eq. (13.9) clearly exhibit marginal densities that are characterized as follows.

$$\varrho(t, x) = \overset{\circ}{\varrho}(x) \qquad \forall\, t > 0\,, \qquad x \in \mathbb{D}\,, \tag{13.12}$$

and

$$\varsigma(t, v) = \overset{\circ}{\varsigma}(v) \qquad \forall\, t > 0\,, \qquad v \in \mathbb{V}\,, \tag{13.13}$$

where $\overset{\circ}{\varrho}(x)$ and $\overset{\circ}{\varsigma}(v)$ are the corresponding initial values

$$\overset{\circ}{\varrho}(x) = \int_{\mathbb{V}} \overset{\circ}{f}(x, w)\, \mathrm{d}w\,, \qquad \overset{\circ}{\varsigma}(v) = \int_{\mathbb{D}} \overset{\circ}{f}(y, v)\, \mathrm{d}y\,.$$

Then we notice that Eq. (13.9) results in

$$\frac{\partial}{\partial t} f(t, x, v) = \overset{\circ}{\varrho}(x)\, \overset{\circ}{\varsigma}(v) - f(t, x, v)\,, \tag{13.14}$$

and

$$f(t, x, v) = \overset{\circ}{f}(x, v) e^{-t} + \overset{\circ}{\varrho}(x)\, \overset{\circ}{\varsigma}(v) \left(1 - e^{-t}\right)\,, \tag{13.15}$$

for every $t > 0$, $x \in \mathbb{D}$ and $v \in \mathbb{V}$. Consequently, we get; see Ref. [83],

Corollary 10 *Given a solution $f = f(t)$ to Eq. (13.9) that is a probability density. The marginal probability densities $\varrho(t,x)$ and $\varsigma(t,v)$ are conserved in $t > 0$, that is, Eqs. (13.12), (13.13) are satisfied and*

$$|f(t,x,v) - \overset{\circ}{\varrho}(x)\,\overset{\circ}{\varsigma}(v)| \leq \mathfrak{c}\, e^{-t}\,,$$

where the constant "$\mathfrak{c}$" depends on $\overset{\circ}{f}$.

This indicates that, regardless of its appealing mathematical structure, Eq. (13.9) can play a rather restricted role in effectively characterizing the tendon recovery process. In this chapter, see Eq. (13.30), we propose a more complex and possibly more adequate model, still with $\gamma = 1$, that is, a system of 3 equations, one of which is of type Eq. (13.10) with $\gamma = 1$.

As discussed in the preceding chapters, the scenario where $\gamma > 1$ presents a significantly higher level of complexity and has the potential to produce more intriguing results, particularly when we keep in mind the interpretation of blow–ups as a description of tendon recovery. Generally, similar to the results in Chapter 4, we can expect only a local existence result; see Ref. [83].

Theorem 35 *Let*

- $\gamma > 1$;
- $a \in L^\infty\left((\mathbb{D}\times\mathbb{V})^2\right)$ *and* $a \geq 0$;
- $\overset{\circ}{f}$ *be a probability density such that* $\overset{\circ}{f} \in L^\infty(\mathbb{D}\times\mathbb{V})$.

Then there exist $\tau > 0$ and a unique function $f = f(t)$ in $L^\infty(\mathbb{D}\times\mathbb{V})$ that is a solution to Eq. (13.10), corresponding to the initial datum $\overset{\circ}{f}$, for $t \in [0,\tau]$. The solution $f = f(t)$ is a probability density and the marginal density ϱ is conserved in the following sense

$$\varrho(t,x) = \overset{\circ}{\varrho}(x)\,, \qquad x \in \mathbb{D}\,, \tag{13.16}$$

for each $t \in [0,\tau]$. Moreover, if both a and $\overset{\circ}{f}$ are continuous, then $f = f(t)$ is a continuous function on $\mathbb{D}\times\mathbb{V}$ for each $t \in [0,\tau]$.

Proof. The proof follows in a standard way—see Ref. [83]. By direct integration of Eq. (13.10) with respect to $v \in \mathbb{V}$ we obtain Eq. (13.16). ■ QED

The principle of conservation of marginal density $\varrho = \varrho(t,x)$, as stated in Eq. (13.16), is explained by the observation that the model, given by Eq. (13.10), does not describe, according to its biological meaning, factors such as movement, proliferation, and destruction.

Let $a = 1$ and

$$\varsigma^{[\gamma]}(t,v) = \int_{\mathbb{D}} f^\gamma(t,y,v)\,\mathrm{d}y$$

and as before ς denotes $\varsigma^{[1]}$. We can rewrite Eq. (13.10) in the following form.

$$\frac{\partial}{\partial t} f(t,x,v) = \varsigma^{[\gamma]}(t,v) \int_{\mathbb{V}} f(t,x,w)\,\mathrm{d}w - f(t,x,v) \int_{\mathbb{V}} \varsigma^{[\gamma]}(t,w)\,\mathrm{d}w\,. \tag{13.17}$$

Therefore, we obtain

$$\frac{\partial}{\partial t} f(t,x,v) = \overset{\circ}{\varrho}(x)\varsigma^{[\gamma]}(t,v) - f(t,x,v) \int_{\mathbb{V}} \varsigma^{[\gamma]}(t,w)\,\mathrm{d}w\,. \tag{13.18}$$

Similarly to what was presented in Chapter 4, we now consider

$$F(t) = \int_0^t \int_{\mathbb{V}} \varsigma^{[\gamma]}(t,w)\,\mathrm{d}w\,.$$

We obtain

$$\frac{\partial}{\partial t}\left(f(t,x,v)e^{F(t)}\right) = \overset{\circ}{\varrho}(x)\varsigma^{[\gamma]}(t,v)e^{F(t)}\,. \tag{13.19}$$

From Eq. (13.18) it follows

$$\frac{\partial}{\partial t}\varsigma(t,v) = \varsigma^{[\gamma]}(t,v) - \varsigma(t,v)\,\frac{\mathrm{d}}{\mathrm{d}t}F(t)\,,$$

and

$$\varsigma(t,v)e^{F(t)} = \overset{\circ}{\varsigma}(v) + \int_0^t e^{F(s)}\varsigma^{[\gamma]}(s,v)\,\mathrm{d}s\,.$$

Consequently, we conclude that ς is not conserved, which differs from the scenario examined in Corollary 10. It is evident that the temporal behaviour of ς is characterized by the function $\varsigma^{[\gamma]}$. The application of Eq. (13.19) yields

$$f(t,x,v)e^{F(t)} = \overset{\circ}{f}(x,v) + \overset{\circ}{\varrho}(x) \int_0^t \varsigma^{[\gamma]}(s,v)e^{F(s)}\,\mathrm{d}s\,,$$

and

$$f^{\gamma}(t,x,v)e^{\gamma F(t)} = \left(\overset{\circ}{f}(x,v) + \overset{\circ}{\varrho}(x) \int_0^t \varsigma^{[\gamma]}(s,v)e^{F(s)}\,\mathrm{d}s\right)^{\gamma}.$$

Integrating with respect to x we obtain

$$\varsigma^{[\gamma]}(t,v)e^{\gamma F(t)} = \int_{\mathbb{D}} \left(\overset{\circ}{f}(x,v) + \overset{\circ}{\varrho}(x) \int_0^t \varsigma^{[\gamma]}(s,v)e^{F(s)}\,\mathrm{d}s\right)^{\gamma} \mathrm{d}x\,.$$

Then we have

$$\frac{\varsigma^{[\gamma]}(t,v)\,e^{F(t)}}{\int_{\mathbb{D}} \left(\overset{\circ}{f}(x,v) + \overset{\circ}{\varrho}(x) \int_0^t \varsigma^{[\gamma]}(s,v)e^{F(s)}\,\mathrm{d}s\right)^{\gamma} \mathrm{d}x} = e^{(1-\gamma)\,F(t)}\,.$$

Given $v \in \mathbb{V}$, let

$$H_v(\Xi) = \int_0^{\Xi} \frac{1}{\int_{\mathbb{D}} \left(\overset{\circ}{f}(x,v) + \overset{\circ}{\varrho}(x)\,u\right)^{\gamma} \mathrm{d}x}\,\mathrm{d}u\,.$$

We obtain

$$H_v\left(\int_0^t \varsigma^{[\gamma]}(s,v)e^{F(s)}\,\mathrm{d}s\right)=\int_0^t e^{(1-\gamma)\,F(s)}\,\mathrm{d}s\,,$$

and

$$\int_0^t \varsigma^{[\gamma]}(s,v)e^{F(s)}\,\mathrm{d}s = H_v^{-1}\left(\int_0^t e^{(1-\gamma)\,F(s)}\,\mathrm{d}s\right),$$

where, for any $v\in\mathbb{V}$, H_v^{-1} is the inverse function of H_v and is defined in its maximal domain $[0,\Xi_*(v))$. integrating with respect to v we obtain

$$\int_0^t \frac{\mathrm{d}F}{\mathrm{d}t}(s)\,e^{F(s)}\,\mathrm{d}s=\int_{\mathbb{V}} H_v^{-1}\left(\int_0^t e^{(1-\gamma)\,F(s)}\,\mathrm{d}s\right)\mathrm{d}v\,,$$

and then

$$e^{F(t)}-1=\int_{\mathbb{V}} H_v^{-1}\left(\int_0^t e^{(1-\gamma)\,F(s)}\,\mathrm{d}s\right)\mathrm{d}v\,.$$

It follows

$$1=e^{(1-\gamma)F(t)}\left(1+\int_{\mathbb{V}} H_v^{-1}\left(\int_0^t e^{(1-\gamma)\,F(s)}\,\mathrm{d}s\right)\mathrm{d}v\right)^{\gamma-1}. \tag{13.20}$$

Integrating with respect to t, yields

$$t=\int_0^{\vartheta(t)}\left(1+\int_{\mathbb{V}} H_v^{-1}(\xi)\,\mathrm{d}v\right)^{\gamma-1}\mathrm{d}\xi\,,$$

where

$$\vartheta(t)=\int_0^t e^{(1-\gamma)\,F(s)}\,\mathrm{d}s\,. \tag{13.21}$$

Denoting

$$\Xi_*=\inf_v \Xi_*(v) \tag{13.22}$$

we have

$$\tau=\int_0^{\Xi_*}\left(1+\int_{\mathbb{V}} H_v^{-1}(\xi)\,\mathrm{d}v\right)^{\gamma-1}\mathrm{d}\xi\,, \tag{13.23}$$

where τ can be such that $\tau<\infty$ or $\tau=\infty$.

Keeping this in mind, we can propose the following result; see Ref. [83], assuming for simplicity that the interaction rate is constant as in Chapter 4. The more complicated situation of the changing interaction rate requires further research; cf. Chapter 5.

We have the following result; see Ref. [83], Theorem 4.3.

Theorem 36 *Let*

- $\gamma>1$,

- *a be constant:* $a = 1$,
- $\overset{\circ}{f}$ *be a probability density such that it is a continuous bounded function on* $\mathbb{D} \times \mathbb{V}$,
- $f = f(t)$ *be the unique solution of Eq.* (13.10), *corresponding to the initial datum* $\overset{\circ}{f}$, *given by Theorem 35, in some interval* $[0, \tau]$.

Then the solution f *can be extended to the interval* $[0, \tau_1)$, $\tau_1 > \tau$, *and one of the following conditions holds*

- $\tau_1 < \infty$ *and* τ_1 *is a time of blow-up of the solution* $f = f(t)$,
- $\tau_1 = \infty$ *and the solution exists globally, that is for any* $t > 0$.

In both cases, the marginal density ϱ *is conserved, that is, Eq.* (13.16) *is satisfied, on the interval of time of existence of the solution.*

Remark 3 *The continuity assumption, which can certainly be relaxed, in Theorem 36 serves to define* Ξ_*; *cf. Eq.* (13.22).

Keeping $a = 1$, for ease of notation, we provide calculations for the case $\gamma = 2$. It follows

$$H_v(\Xi) = \int_0^{\Xi} \frac{1}{\int_{\mathbb{D}} \left(\overset{\circ}{f}(x,v) + \overset{\circ}{\varrho}(x)\, u \right)^2 \mathrm{d}x} \,\mathrm{d}u = \int_0^{\Xi} \frac{1}{A_1 u^2 + 2A_2(v) u + A_3(v)} \,\mathrm{d}u \,.$$

where the quantities

$$\begin{aligned} B_1 &= \int_{\mathbb{D}} \left(\overset{\circ}{\varrho}(x) \right)^2 \mathrm{d}x \,, \\ B_2(v) &= \int_{\mathbb{D}} \overset{\circ}{\varrho}(x)\, \overset{\circ}{f}(x,v)\, \mathrm{d}x \,, \\ B_3(v) &= \int_{\mathbb{D}} \left(\overset{\circ}{f}(x,v) \right)^2 \mathrm{d}x \end{aligned}$$

are expressed by the initial data $\overset{\circ}{f}$. Next we define

$$\Upsilon(v) = -\frac{\Delta(v)}{4} \qquad \text{where} \qquad \Delta(v) = 4B_2^2(v) - 4B_1 B_3(v) \,,$$

and assume that

$$\Upsilon(v) > 0 \qquad \forall\, v \in \mathbb{V} \,. \tag{13.24}$$

The Cauchy–Schwarz inequality yields

$$\Delta(v) \leq 0 \,,$$

The explicit formula for $H_v(\Xi)$ has the form

$$H_v(\Xi) = \frac{1}{\sqrt{\Upsilon(v)}} \left(\arctan\left(\frac{B_1 \Xi + B_2(v)}{\sqrt{\Upsilon(v)}} \right) - \arctan\left(\frac{B_2(v)}{\sqrt{\Upsilon(v)}} \right) \right),,$$

and then we obtain the formula for $H_v^{-1} = H_v^{-1}(Y)$

$$H_v^{-1}(Y) = \frac{\sqrt{\Upsilon(v)}}{B_1} \frac{\frac{B_2(v)}{\sqrt{\Upsilon(v)}} + \tan\left(\sqrt{\Upsilon(v)}\,Y\right)}{1 - \frac{B_2(v)}{\sqrt{\Upsilon(v)}} \tan\left(\sqrt{\Upsilon(v)}\,Y\right)} - \frac{\sqrt{\Upsilon(v)} B_2(v)}{B_1} . \tag{13.25}$$

Applying Eq. (13.20) and considering Eq. (13.21), we get the following ordinary differential equation

$$\dot{\vartheta}(t) = \left(1 + \int_{\mathbb{V}} H_v^{-1}\left(\vartheta(t)\right) \mathrm{d}v\right)^{-1} ,$$

that can be rewritten in the form

$$\dot{\vartheta}(t) =$$

$$= \left(1 + \int_{\mathbb{V}} \left(\frac{\sqrt{\Upsilon(v)}}{B_1} \frac{\frac{B_2(v)}{\sqrt{\Upsilon(v)}} + \tan\left(\sqrt{\Upsilon(v)}\,\vartheta(t)\right)}{1 - \frac{B_2(v)}{\sqrt{\Upsilon(v)}} \tan\left(\sqrt{\Upsilon(v)}\,\vartheta(t)\right)} - \frac{\sqrt{\Upsilon(v)}\,B_2(v)}{B_1}\right) \mathrm{d}v\right)^{-1} . \tag{13.26}$$

The initial data are such that

$$\overset{\circ}{\vartheta} = 0 ,$$

by Eq. (13.21). The right–hand side of Eq. (13.26) is positive and thus the function $\vartheta = \vartheta(t)$ increases in some interval $[0, \vartheta_*)$, where

$$\vartheta_* := \lim_{t \uparrow t_*} \vartheta(t) , \tag{13.27}$$

is given by the following condition

$$\sup_{v \in \mathcal{V}} \frac{B_2(v)}{\sqrt{\Upsilon(v)}} \tan\left(\sqrt{\Upsilon(v)}\,\vartheta_*\right) = 1 . \tag{13.28}$$

The condition given by Eq. (13.28) originates from the denominator of the integrand in Eq. (13.26). The function in the denominator is continuous and the singularity appears when it equals zero. We observe that blow–ups can occur if the denominator in Eq. (13.26) exhibits a singularity, which corresponds to the satisfaction of Eq. (13.28). Keeping in mind the previously given interpretation of blow-ups, we can state that, if a subset of orientations that are initially dominant, in the entire set of orientations, is small, then the recovery of the tendon is effectively implemented in a finite time interval. The situation here is analogous to that we observe in the previous chapters; see Chapter 4.

We denote the right–hand side of Eq. (13.26) by $\Phi = \Phi(\vartheta)$, that is

$$\Phi(\vartheta_*) = \lim_{\vartheta \uparrow \vartheta_*} \Phi(\vartheta) . \tag{13.29}$$

In a spirit similar to Theorem 6, the following result holds; see Ref. [83], Theorem 4.4.

Theorem 37 *Let*

- $\gamma = 2$;

- $a = 1$;
- *The probability density* $\overset{\circ}{f}$ *be a continuous and bounded function on* $\mathbb{D} \times \mathbb{V}$*;*
- $f = f(t)$ *be the unique solution of Eq.* (13.10) *corresponding to the initial datum* $\overset{\circ}{f}$.

We have

1. *If* $\Phi(\vartheta_*) > 0$, *then there is a blow–up of* f *in a finite time* $\tau > 0$*;*
2. *If* $\Phi(\vartheta_*) = 0$, *then*

   (a) *if* $\int_0^{\vartheta_*} \frac{1}{\Phi(\vartheta)} \, \mathrm{d}\vartheta < \infty$ *then there is a blow–up of* f *in a finite time* $\tau > 0$*;*

   (b) *if* $\int_0^{\vartheta_*} \frac{1}{\Phi(\vartheta)} \, \mathrm{d}\vartheta = \infty$ *then* f *is a global solution.*

Remark 4 *In Items* 1 *and* 2(*a*) *of Theorem 37, the blow–up time* $\tau < \infty$ *is defined by the formula*

$$\tau = \int_0^{\vartheta_*} \left(1 + \int_{\mathbb{V}} \left(\frac{\sqrt{\Upsilon(v)}}{B_1} \frac{\frac{B_2(v)}{\sqrt{\Upsilon(v)}} + \tan\left(\sqrt{\Upsilon(v)}\,\vartheta\right)}{1 - \frac{B_2(v)}{\sqrt{\Upsilon(v)}} \tan\left(\sqrt{\Upsilon(v)}\,\vartheta\right)} - \frac{\sqrt{\Upsilon(v)}\, B_2(v)}{B_1} \right) \mathrm{d}v \right) \mathrm{d}\vartheta \,,$$

whereas in Item 2(*b*) *we have* $\tau = \infty$.

Remark 5 *By Eq.* (13.18) *we have*

$$\frac{\partial}{\partial t} \varsigma(t, v) = \varsigma^{[\gamma]}(t, v) - \varsigma(t, v) \frac{\mathrm{d}}{\mathrm{d}t} F(t) \,,$$

and

$$\varsigma(t, v) e^{F(t)} = \overset{\circ}{\varsigma}(v) + \int_0^t e^{F(s)} \varsigma^{[\gamma]}(s, v) \, \mathrm{d}s \,.$$

Therefore, $\varsigma(t, v)$ *is not conserved, in contrast to the case in Corollary 10 even in the case of factorized initial data*

$$\overset{\circ}{f}(x, v) = \overset{\circ}{\varrho}(x) \, \overset{\circ}{\varsigma}(v) \,.$$

It corresponds to $\Upsilon = 0$ *contrary to Eq.* (13.24)*. Therefore, the solution* $f(t, x, v) = \overset{\circ}{\varrho}(x) \varsigma(t, v)$ *relates to an equation analogous to that in Chapter 4. For these reasons, blow–ups can occur like in Chapter 4.*

Theorem 37 shows that when $\gamma = 2$ and the interaction rate is constant $a = 1$, the nature of the solutions is essentially determined by the initial data. Specifically, the solutions can either be defined over an infinite time interval, that is, global solutions exist, or can have a blow–up in a finite time interval. This latter phenomenon may be linked to successful recovery, as previously discussed. The emergence of blow–ups in solutions within a finite time interval, shown rigorously for certain initial data and illustrated through numerical simulations; see Ref. [83], can be associated with the alignment of collagen fibres occurring within a finite interval of time. This behaviour can be linked to the possibility of fully healing from the injury without the development of scars. The significance of this aspect of the model is important and warrants thorough investigation in future studies, particularly since scarring is a common outcome following tendon injuries and significantly affects the condition of the tendon.

It should be remembered that the analysis was provided for Eq. (13.10) in the simplified situation where a is a constant, $a = 1$ and $\gamma = 2$. This is an essential simplification that significantly simplifies mathematical analysis, but it can reasonably be hypothesized that it encapsulates the fundamental characteristics of the behaviour of the potential solution in a general context. In Chapters 5 and 6, in a simpler situation, it was visible that the rejection of the assumption on a constant kernel leads to complicated considerations and requires other assumptions that are not always justified by applications. Mathematical analysis and simulations, cf. Ref. [83], have shown the substantial impact of initial data on the final structure of collagen fibres. This could be an essential indication for surgical intervention treatment from a medical point of view. Although simplified, the model demonstrated interesting characteristics from both mathematical and modelling points of view.

Based on a similar kinetic description as before of the orientation of the fibre bundles, by including local alignment interactions between the fibres, Carrillo et al. Ref. [50] proposed a model of collagen fibre rearrangement during the tendon healing process. Using the methodology of asymptotic grazing collision limits; Ref. [50], derived a model—a nonlinear non–local Fokker–Planck type equation with a decreased number of parameters, making parameter estimation easier. The model of tendon healing in the mathematical structure of Ref. [50] is similar to kinetic models of alignment or consensus processes rather than to models of wound healing. The reader is referred to a short overview of the related models in Ref. [50].

13.2.2 Role of fibroblasts

Reference [135] is a continuation of Ref. [83]. A model is proposed that describes the reconstruction of the parallel structure of collagen fibres, includes the reparative role of tendon fibroblasts, and takes into account their degree of maturity. Tendon cells are mainly tendon–specific fibroblasts; see Refs. [191, 219]. Keeping in mind that cell movement is a process much faster than collagen remodelling, and referring to the healing phase after the relocation of most of the cells, we can neglect the movement of cells. Unlike the case considered above, the model assumes that $\gamma = 1$, which, as we shall see, eliminates the possibility of blow-ups—cf. Theorem 34 above and see Theorem 38 below.

In the previous section, blow–ups were interpreted as the healing of tendons in a finite time. The model in Ref. [135] may be interesting as a description of how the healthy state is achieved. Thus, the modelling strategy is now different from the one in the previous section. It involves examining the ways of reaching a healthy tendon depending on the role of tendon fibroblasts.

The model reads

$$\begin{aligned}
\tfrac{\partial}{\partial t} f(t,x,v) &= \mathfrak{G}[f,h_1](t,x)\,, \\
\tfrac{\partial}{\partial t} h_0(t,x) &= \beta(x) + \xi_0\Big(\mathfrak{v}(f)(t,x)\Big)h_0(t,x) - \Big(\alpha + \nu_0\Big)\, h_0(t,x)\,, \\
\tfrac{\partial}{\partial t} h_1(t,x) &= \alpha\, h_0(t,x) - \nu_1 h_1(t,x)\,,
\end{aligned} \tag{13.30}$$

where

$$\mathfrak{G}[f,h_1](x,v) =$$

$$= \int_{\mathbb{D}}\int_{\mathbb{V}} a\Big(y,v;x,w;\ \Lambda\big(h_1(x)\big)\Big)\, f(y,v)\, f(x,w)\, \mathrm{d}w\, \mathrm{d}y \tag{13.31}$$

$$- \quad f(x,v)\int_{\mathbb{D}}\int_{\mathbb{V}} a\Big(y,w;x,v;\ \Lambda\big(h_1(x)\big)\Big)\, f(y,w)\, \mathrm{d}w\, \mathrm{d}y\,.$$

The function $f = f(t,x,v)$ describes the probability density to find a collagen fibre in the instant of time $t > 0$ at the point $x \in \mathbb{D}$ and with orientation $v \in \mathbb{V}$. Here $\mathbb{D}$ and $\mathbb{V}$ are domains in $\mathbb{R}^{\mathfrak{n}}$ and $\mathbb{R}^{l}$, respectively, for some integers $\mathfrak{n}$ and l. To avoid technical difficulties, we assume that $\mathbb{D}$ is identified with the $\mathfrak{n}$–dimensional torus, $\mathbb{D} = \mathbb{T}^{\mathfrak{n}}$. The variable v, as before, describes the direction of collagen fibres. We can interpret v in terms of angle and assume that $\mathbb{V} = \left[-\frac{\pi}{2}, \frac{\pi}{2}\right]$. The function $h_0 = h_0(t,x)$ and $h_1 = h_1(t,x)$ are the densities of immature fibroblasts (tenoblasts) and mature fibroblasts (tenocytes), respectively, at time $t > 0$, at the point $x \in \mathbb{D}$. The constant parameters $\alpha > 0$, $\nu_0 > 0$ and $\nu_1 > 0$ are the maturation rate of tenoblasts, the mortality rate of tenoblasts, and the mortality rate of tenocytes, respectively. The function $\beta = \beta(x)$ describes an inflow of tenoblasts. In the model Eq. (13.30) we do not allow movement of tenoblasts, so it is natural to assume that β is a nonnegative constant. In a more general situation where the movement of tenoblasts is included; see Eq. (13.45); more natural is to assume that β has a support in some subdomain of $\mathbb{D}$.

The kernel a, which defines the turning rate, depends on the function Λ—the efficiency of the remodelling process that depends on $h_1(t,x)$.

Consistent with the modern perspective that tendon damage involves distortion of the parallel alignment of collagen fibres, we can use the second-moment $\mathfrak{v}(f)(t,x)$ with respect to the variable v to model the damage. We introduce

$$\mathfrak{v}(f)(t,x) = \frac{1}{\varrho(t,x)} \int_{\mathbb{V}} \Big(v - \mathfrak{m}(t,x)\Big)^2 f(t,x,v)\, \mathrm{d}v\,, \tag{13.32}$$

where

$$\varrho(t,x) = \int_{\mathbb{V}} f(t,x,v)\, \mathrm{d}v\,, \qquad \mathfrak{m}(t,x) = \frac{1}{\varrho(t,x)} \int_{\mathbb{V}} v f(t,x,v)\, \mathrm{d}v\,.$$

The function ξ_0 that depends on $\mathfrak{v}(f)(t,x)$ is the production of tenoblasts; in fact, the tendon injury activates cell proliferation.

The global existence and uniqueness theory for Eq. (13.30) is standard under the assumptions of the theorem below—Theorem 38.

Let $X = L^\infty\Big(\mathbb{V}\times\mathbb{D}\Big) \times L^\infty(\mathbb{D}) \times L^\infty(\mathbb{D})$ be the space equipped with the norm

$$\|(f,h_0,h_1)\| = \max\{\|f\|_\infty, \|h_0\|_\infty, \|h_1\|_\infty\}\,,$$

where $\|\cdot\|_\infty$ is the standard norm in $L^\infty\Big(\mathbb{V}\times\mathbb{D}\Big)$ or $L^\infty(\mathbb{D})$—the meaning depends on the context.

The cone in X of functions (f, c_0, c_1) such that f, c_0 and c_1 are non–negative is denoted by X_+.

We have the following result; see Ref. [135],

Theorem 38 *Let*

- *the kernel* a *be a measurable bounded non–negative function that is globally Lipschitz continuous with respect to the last variable* Λ,
- $\Lambda : \mathbb{R}^1_+ \to \mathbb{R}^1_+$ *and* $\xi_0 : \mathbb{R}^1_+ \to \mathbb{R}^1_+$ *be globally Lipschitz continuous functions;*
- *the function* β *is a measurable bounded non–negative function on* $\mathbb{D}$,
- $\alpha > 0$, $\nu_0 > 0$ *and* $\nu_1 > 0$ *be given parameters,*
- $(\overset{\circ}{f}, \overset{\circ}{h}_0, \overset{\circ}{h}_1) \in X_+$ *be such that* $\overset{\circ}{f}$ *be a probability density on* $\mathbb{D} \times \mathbb{V}$ *and*

$$\overset{\circ}{\varrho}(x) \equiv \int_{\mathbb{V}} \overset{\circ}{f}(x,v)\,\mathrm{d}v > 0 \quad \text{for all } x \in \mathbb{D}\,. \tag{13.33}$$

Then, for any $\tau > 0$, *there exists a unique solution* (f, h_0, h_1) *of Eq.* (13.30), *corresponding to the initial data* $(\overset{\circ}{f}, \overset{\circ}{h}_0, \overset{\circ}{h}_1) \in X_+$, *in* X *on interval* $[0,\tau]$ *and is such that*

- $f = f(t)$ *is a probability density on* $\mathbb{D} \times \mathbb{V}$, *for all* $t \in [0,\tau]$,
- $\varrho(t,x) \equiv \int_{\mathbb{V}} f(t,x,v)\,\mathrm{d}v = \overset{\circ}{\varrho}(x)$ *for all* $x \in \mathbb{D}$ *and for all* $t \in [0,\tau]$,
- $h_0(t) \geq 0$ *and* $h_1(t) \geq 0$ *for all* $t \in [0,\tau]$.

Proof. The operator $\mathfrak{G}$, given by the right–hand side of the first equation of Eq. (13.30), satisfies the following conservative property; cf. Chapter 3,

$$\begin{aligned}
&\int_{\mathbb{V}} \phi(v)\mathfrak{G}[f](x,v)\,\mathrm{d}v = \\
&= \int_{\mathbb{V}}\int_{\mathbb{V}}\int_{\mathbb{D}} \phi(v)\, a\Big(y,v;x,w;\, \Lambda\big(h_1(x)\big)\Big)\, f(y,v)f(x,w)\,\mathrm{d}y\,\mathrm{d}w\,\mathrm{d}v \\
&- \int_{\mathbb{V}}\int_{\mathbb{V}}\int_{\mathbb{D}} \phi(v)\, a\Big(y,w;x,v;\, \Lambda\big(h_1(x)\big)\Big)\, f(x,v)\, f(y,w)\,\mathrm{d}y\,\mathrm{d}w\,\mathrm{d}v = \\
&= \int_{\mathbb{V}}\int_{\mathbb{V}}\int_{\mathbb{D}} \big(\phi(v) - \phi(w)\big)\, a\Big(y,v;x,w;\, \Lambda\big(h_1(x)\big)\Big) \times \\
&\times\ f(y,v)f(x,w)\,\mathrm{d}y\,\mathrm{d}w\,\mathrm{d}v\,.
\end{aligned} \tag{13.34}$$

In particular, taking $\phi \equiv 1$, we obtain

$$\int_{\mathbb{V}} \mathfrak{G}[f](t,x,v)\,\mathrm{d}v = 0\,, \tag{13.35}$$

for $t > 0$ and $x \in \mathbb{D}$ and any integrable function f.

Then we observe, by Eq. (13.35), that the solution *a priori* satisfies

$$\varrho(t,x) = \overset{\circ}{\varrho}(x) \qquad x \in \mathbb{D}\,, \qquad t \geq 0. \tag{13.36}$$

Therefore, Eq. (13.32) *a priori* makes sense. The solution is *a priori* non–negative. It is easy to see that the operator, defined by the right–hand side of Eq. (13.30), is (locally) Lipschitz continuous in X. Therefore, there exists $\tau_1 > 0$ and the (local–in–time) solution (f, c_0, c_1) of Eq. (13.30) for $t \in [0, \tau_1]$. It is easy to see that f is a probability density and f is bounded on each compact time interval. In particular $f(t) \geq 0$. Then,

$$\|f(t)\|_{L^1(\mathbb{D}\times\mathbb{V})} = \| \overset{\circ}{f} \|_{L^1(\mathbb{D}\times\mathbb{V})} = 1\,, \tag{13.37}$$

and keeping in mind that $\mathbb{V} = [-\frac{\pi}{2}, \frac{\pi}{2}]$, we have

$$0 \leq \mathfrak{v}[f](t,x) \leq \left(\frac{\pi}{2}\right)^2 . \tag{13.38}$$

Moreover we have

$$\|f(t)\|_\infty \leq \mathfrak{c}_1\,, \qquad t \in [0, \tau_1]\,, \tag{13.39}$$

where the constant denoted by $\mathfrak{c}_1$ depends on τ_1. Finally, we obtain

$$\sup_{t\in[0,\tau]} \|(f, h_0, h_1)\| \leq \mathfrak{c}\,, \tag{13.40}$$

where $\mathfrak{c}$ is a constant depending on τ, but bounded for any bounded τ. This ends the proof. ■ QED

The model Eq. (13.30), in its generality, can describe various scenarios of the behaviour of collagen fibres, including recovery and scarring formation. Therefore, in order to describe the recovery process, it is worth formulating mathematical conditions that guarantee recovery. We are going to propose a condition (Assumption 17 below) that implies

$$\int_{\mathbb{V}} |v|^2\, \mathfrak{G}[f](t,x,v)\, dv \leq 0\,, \tag{13.41}$$

for $t > 0$ and $x \in \mathbb{D}$ and any non–negative integrable function f. The condition (13.41) reflects the dissipative property of the interaction operator that leads to tendon healing expressed in terms of $\mathfrak{V}(f)$ tending to 0 for $t \to \infty$

Assumption 17 *Let*

$$a\Big(y, v; x, w;\, \Lambda\,(h_1(t,x))\Big) \;=\; h_1(t,x)\,\chi\Big(|v| < |w|\Big)\,. \tag{13.42}$$

where χ *is given by Eq.* (3.1).

It is important to note that, in an entirely healthy tendon, the collagen fibres align along the tendon's longitudinal axis. The Assumption 17 means that in the presence of tenocytes, the repair process always occurs, which in modelling terms means the change from orientation w to orientation v for $|v| < |w|$.

The assumption 17 implies that

$$\mathfrak{v}[f](t,x) \quad\to\quad 0 \qquad \text{as} \quad t \to \infty\,, \qquad \forall\, x \in \mathbb{D}\,. \tag{13.43}$$

The assumption 17 is quite restrictive. In fact, it implies not only Eq. (13.41) but generally

$$\int_{\mathbb{V}} |v|^i\, \mathfrak{G}[f](t,x,v)\, dv \leq 0\,, \tag{13.44}$$

for any even integer $i \geq 1$ and any non–negative integrable function f.

Remark 6 *Assuming that*

$$\xi_0(0) < \alpha + \nu_0 \,,$$

but

$$\beta(x) + \xi_0(\mathfrak{v}[f](0,x))\, \overset{\circ}{h}_0(x) > (\alpha + \nu_0)\; \overset{\circ}{h}_0\,(x)\,,$$

we obtain that $h_0(\,\cdot\,,x)$ *increases for some time* $\tilde{t}_1 > 0$ *such that*

$$\beta(x) + \xi_0(\mathfrak{v}[f](\tilde{t}_1,x))h_0(\tilde{t}_1,x) = (\alpha + \nu_0)h_0(\tilde{t}_1,x)\,.$$

After reaching some $\tilde{t}_2 \geq \tilde{t}_1$ $h_0(\,\cdot\,,x)$ *starts to decrease tending to* $\frac{\beta(x)}{\alpha+\nu_0-\xi_0(0)}$. *The dynamics of tendon cells express the actions of tenoblasts and tenocytes during the healing phase. Specifically, the initial increase of tenoblasts is due to the need for increased collagen production, which is subsequently followed by a decrease in tenoblasts density.*

13.2.3 Non–local model of tendon healing process

Keeping in mind the results of Section 12.3 we can come back to the modelling of the tendon healing process considered in Section 13.2 and propose a purely non–local model, with respect to the spatial variable, consisting of integro–differential equations describing the reconstruction of collagen fibres.

We take into account that tendon injury activates not only cell proliferation but also tendon fibroblast movement toward the highest level of tendon damage that could be measured by the second moment Eq. (13.32). Since movement affects only tenoblasts and not tenocytes, the appropriate terms describing movement are only included in the equation for h_0. Therefore, the non–local system reads

$$\begin{aligned}
\frac{\partial}{\partial t} f(t,x,v) &= \mathfrak{G}[f,h_1](t,x,v)\,,\\
\frac{\partial}{\partial t} h_0(t,x) &= \xi_0\Big(\mathfrak{v}(f)(t,x)\Big)h_0(t,x) - (\alpha+\nu_0)h_0(t,x) + \\
&\quad + \ \mathcal{G}_{DL}[h_0](t,x) - \mathcal{G}_T[h_0\,,\,\mathfrak{v}(f)](t,x)\,,\\
\frac{\partial}{\partial t} h_1(t,x) &= \lambda h_0(t,x) - \nu_1 h_1(t,x).
\end{aligned} \tag{13.45}$$

Here, the notation of Section 13.2 is used, and in particular, the operator $\mathfrak{G}$ is given by Eq. (13.31). Moreover, $\mathcal{G}_{DL}$ is the linear non–local operator defined by Eq. (12.25) and $\mathcal{G}_T$ is the non–local taxis operator defined by Eq. (12.26).

Using the notation before Theorem 38, we can formulate the analogous global–in–time existence–uniqueness result for the system Eq. (13.45). Both the formulation of the theorem and the proof are analogous to Theorem 38.

13.3 DNA denaturation

13.3.1 DNA: Microscopic model

The process in which DNA undergoes thermal denaturation, known as the DNA melting process, involves the separation of base pairs due to elevated temperatures. It was considered, e.g. in Refs. [10, 176, 177]; see also references therein. Reference [66] deals with the process of DNA melting on a microscopic scale in the framework of Chapter 9. For each base pair the stretching parameter, a microscopic variable $v \in [0, \infty)$ is considered to reflect the separation between complementary bases. That is to say, the model pertains to the states, which refer to the distances between two bases in each individual base pair of DNA. The model has a microscopic and consequently stochastic nature and is proposed in a similar spirit to the linear integro–differential equations in Chapter 9. In a manner akin to Chapter 9, the model defines the dynamics of the probability density linked to the states of every base pair. Due to potential variations in temperature over time, the equation has a non–autonomous structure with respect to the time variable $t \geq 0$, unlike the models presented in Chapter 9. This difference is particularly significant compared to the Markov structures discussed in Chapter 9. The model refers to both weaker double hydrogen bonds, that is, A–T, or T–A, as well as stronger triple hydrogen bonds, that is, G-C, or C-G. It also incorporates the stacking interactions between close base pairs. According to numerical simulations in Ref. [66], S–shaped melting curves of DNA are observed and the formation of bubbles appears, exactly like in the experimental observations such as those described in, e.g. Refs. [176, 177]. A mesoscopic (kinetic) model for DNA denaturation was previously introduced in Ref. [74]. However, the mesoscopic model lacks comprehensive details on every base pair as a result of its mesoscopic nature. Alternatively, Ref. [66] introduces a more comprehensive model that incorporates the time evolution of the probability distribution for all hydrogen bonds.

The model proposed in Ref. [66] takes into account two categories of bonds, weaker and stronger, and the direct dependence of changes in temperature. Base pairs A–T (or T–A) and C–G (or G–T) are identified by the discrete variable n, which takes values from 1 to N, $n \in \{1, 2, \dots, N\}$ and N is a number of all base pairs ("bonds"). For each pair of bases n, the continuous variable $v_n \in [0, \infty)$ defines the distance between the two corresponding bases. We refer to v_n as the stretching parameter for the base pair n. The microscopic state of each base pair n is therefore defined by the variable v_n. The discrete variable is associated with one of two subsets of bonds: $\mathbb{J}_2$ corresponding to the two hydrogen bonds, linking A and T, and $\mathbb{J}_3$ corresponding to the three hydrogen bonds, that is, between C and G.

$$\mathbb{J}_2 \cup \mathbb{J}_3 = \mathbb{J}, \qquad \mathbb{J}_2 \cap \mathbb{J}_3 = \emptyset\,.$$

We assume that 3–hydrogen bonds are more resistant to high temperature than 2–hydrogen bonds. The model describes the time evolution of the probability density

$$f = f(t, v_1, \dots, v_N)$$

that defines the distribution of states $v_1, \dots v_N$ of all N bonds.

In Reference [66] the following linear and not necessarily time–autonomous, operators were proposed

- Λ is the term describing stochastic changes of variables $v_1, \dots, v_N$ due to the interactions between two neighbourhood base pairs;

- Γ_T is the term describing stochastic change of variables $v_1, \dots, v_N$ due to high temperature.

The linear integro–differential equation, in the spirit of Chapter 9, reads

$$\frac{\partial}{\partial t} f = \Lambda f + \Gamma_T f\,, \tag{13.46}$$

where $t \geq 0$ and $v_1, \ldots, v_N \in [0, \infty)$,

$$\Lambda f(t, v_1, \ldots, v_N) =$$

$$= \sum_{n=1}^{N-1} \left(\int_0^\infty A_n(v_n; w, v_{n+1}) a(w, v_{n+1}) f(v_1, \ldots, v_{n-1}, w, v_{n+1}, \ldots, v_N)\, dw \right.$$

$$\left. - \quad a(v_n, v_{n+1}) f(t, v_1, \ldots, v_N) \right) +$$

$$+ \sum_{n=2}^{N} \left(\int_0^\infty A_n(v_n; w, v_{n-1}) a(w, v_{n-1}) f(v_1, \ldots, v_{n-1}, w, v_{n+1}, \ldots, v_N)\, dw \right.$$

$$\left. - \quad a(v_n, v_{n-1}) f(t, v_1, \ldots, v_N) \right),$$

and

$$\Gamma_T f(t, v_1, \ldots, v_N) =$$

$$= \sum_{n=1}^{N} \left(\int_0^\infty B_T(v_n; w) b_n(w) f(t, v_1, \ldots, v_{n-1}, w, v_{n+1}, \ldots, v_N)\, dw \right.$$

$$\left. - \quad b_n(v_n) f(t, v_1, \ldots, v_N) \right).$$

Based on the probabilistic interpretation, cf. Chapter 9, we assume that

$$A_n(v; w, w^*) \geq 0, \qquad \int_0^\infty A_n(v; w, w^*)\, dv = 1, \tag{13.47}$$

for $v, w, w^* \in (0, \infty)$, and

$$B_T(v; w) \geq 0, \qquad \int_0^\infty B_T(v; w)\, dv = 1, \tag{13.48}$$

for $v, w \in (0, \infty)$.

As previously stated, the functions A_n and a define the interactions occurring between the nearest–neighbour bonds. We assume; see Ref. [66]

$$a(v, w) = \alpha(w). \tag{13.49}$$

The probability of increasing the stretching parameter of a bond is higher if the stretching parameters of the nearest–neighbour bonds are greater. This is reflected by a measurable positive function α with the following natural form; see Ref. [66],

$$\alpha(w) = \begin{cases} \alpha_0(1 + w) & \text{if} \quad v, w \in [0, R] \\ \alpha_0(1 + R) & \text{if} \quad v \in [0, \infty), w > R \\ \alpha_0 & \text{if} \quad v > R, w \in [0, R], \end{cases} \tag{13.50}$$

where α_0 and R are given positive constants. The function α is bounded. The bonds at the end of the chain, that is, for $n = 1$ and $n = N$, are distinct from the others, $n = 2, \ldots, N-1$, since they have only one neighbouring bond.

The probability of stretching the state increases as the temperature increases. Thus, it is justified the requirement that B_T depends on temperature T in such way that

$$\int_0^\infty a(v) B_T(v; w)\, dv = \gamma_T\, a(w) \qquad \text{for a.a. } w \in (0, \infty)\,, \tag{13.51}$$

where $\gamma_T = 1$ for $T \leq T_0$, for some T_0 and γ_T is an increasing function of T for $T \geq T_0$ Moreover, $a = a(v)$ is a measurable increasing function of v. If $a(v) = v$, we can note that B_T must be singular in $w = 0$. We assume that T can change with time t, cf. Ref. [218]. Hence, γ_T must be a function of time. In this scenario, we assume that

$$\gamma_T \, : \, [0, \infty) \to [1, \infty)$$

is a continuous function of time t. Moreover,

$$b_n(v) = b_n = \begin{cases} b^{(2)} & \text{if} \quad n \in \mathbb{J}_2 \\ b^{(3)} & \text{if} \quad n \in \mathbb{J}_3\,, \end{cases} \tag{13.52}$$

where $b^{(2)}$ and $b^{(3)}$ are non–negative constants such that $0 \leq b^{(3)} \leq b^{(2)} < \infty$ reflecting the fact that the 3–hydrogen bonds are more heat–resistant than the 2–hydrogen bonds.

Under Assumptions (13.47)—(13.51) together with the continuity assumption of B_T on time variable t, the operator $\mathcal{A} + \mathcal{B}_T$ on the right–hand side of Eq. (13.46), that can depend on time t, is Lipschitz–continuous in the space $L^1_{(N)}$, where we use the notation of Chapter 9, on any interval of time $[0, \tau]$, where $\tau > 0$ is arbitrary but fixed. Then there exists a unique solution and, in a standard way, we can show that the solution $f = f(t)$ is nonnegative for nonnegative initial data $\overset{\circ}{f}$ and the L^1–norm is preserved

$$\| f(t) \|_{L^1_{(N)}} = \| \overset{\circ}{f} \|_{L^1_{(N)}} = 1\,, \qquad \text{for } t \in [0, \tau]\,. \tag{13.53}$$

If T does not depend on the time variable t, the Cauchy problem for Eq. (13.46) has the unique solution $f = f(t)$ given by a uniformly continuous semigroup, like in Chapter 9,

$$f(t) = e^{t\,(\mathcal{A} + \mathcal{B}_T)}\, \overset{\circ}{f}$$

in $L^1_{(N)}$ for all $t \geq 0$. Therefore, $\left(e^{t\,(\mathcal{A} + \mathcal{B}_T)}\right)_{t \geq 0}$ is a uniformly continuous semigroup of Markov operators; that is, a uniformly continuous stochastic semigroup. The semigroup is a partially integral Markov semigroup, and its asymptotic behaviour can be described; see Section 9.2. As in Section 9.2, the following assumption can be proposed; see Ref. [66]

Assumption 18 *Let at least one of the following two conditions be satisfied*

- *a is non–zero and*

$$A_n(v; w, w^*) a(w, w^*) \geq g(v) \qquad \forall\, v\,,\, w\,,\, w^* \in [0\,,\, \infty)\,, \tag{13.54}$$

- *b_n are non–zero and*

$$B_T(v, w) b_n(w) \geq g(u) \qquad \forall\, v\,,\, w \in [0\,,\, \infty)\,, \tag{13.55}$$

for a measurable nonnegative function g *on* $[0,\infty)$ *such that*

$$\int_0^\infty g(v)\,\mathrm{d}v > 0\,. \tag{13.56}$$

Assumption 18 yields, cf. Section 9.2, that

$$\ell(v_1,\dots,v_N) = \prod_{i=1}^N g(v_i)\,,$$

is such that $\ell \in L^1_{(N)}$,

$$\int_{[0,\infty[^N} \ell(v_1,\dots,v_N)\,\mathrm{d}v_1\dots\mathrm{d}v_N > 0\,, \tag{13.57}$$

and it is a lower function in the sense of A. Lasota and J. Yorke [149], see Section 9.2. Therefore, it follows that the semigroup $e^{t\,\left(\mathcal{A}+\mathcal{B}_T\right)}$ is asymptotically stable.

13.3.2 DNA: Symmetrisation

In this section, following Section 4 in Ref. [66], we are dealing with a symmetrized variant of Eq. (13.46) that is

$$\frac{\partial}{\partial t} f = \overline{\Lambda} f + \overline{\Gamma} f\,, \tag{13.58}$$

where $t \geq 0,\ v_1\,,\,\dots\,,\,v_N\ \in [0\,,\,\infty)\,,$

$$\overline{\Lambda} f(t,v_1,\dots,v_N) = \sum_{\substack{1\leq n,m\leq N\\ |n-m|=1\ (\mathrm{mod}\ N-2)}} \left(\int_0^\infty A(v_n;w,v_m)a(w,v_m)\,\times \right.$$

$$\left. \times\quad f\,(v_1\,,\,\dots\,,v_{n-1},w,v_{n+1},\dots\,,v_N)\,\mathrm{d}w - a(v_n,v_m)f\,(t,v_1\,,\,\dots\,,v_N) \right),$$

and

$$\overline{\Gamma} f(t,v_1,\dots,v_N) \;=$$

$$= \sum_{n=1}^N \left(\int_0^\infty B(v_n;w)b(w)f\,(t,v_1\,,\,\dots\,,v_{n-1},w,v_{n+1},\dots\,,v_N)\,\mathrm{d}w \right.$$

$$\left. -\quad b(v_n)f\,(t,v_1\,,\,\dots\,,v_N) \right),$$

where $A = A_n$ and $b = b_n$ are independent of n, and $B = B_T$ is independent of T. As previously, $\left(e^{t\,(\overline{\Lambda}+\overline{\Gamma})}\right)_{t\geq 0}$ is a uniformly continuous semigroup of Markov operators, a uniformly continuous stochastic semigroup. Its stability may be stated as before.

It is clear that the solution $f(t) = f(t,v_1,\dots,v_N)$ to Eq. (13.58) is not necessarily a symmetric function with respect to its variables $v_1,\dots,v_N$. Despite this, there is some biological argument for justifying the symmetrical scenario, see Refs. [66] and [194].

For any $f = f(t, v_1, \dots, v_N)$, let $\widetilde{f} = \widetilde{f}(v_1, \dots, v_N)$ be given by

$$\widetilde{f}(v_1, \dots, v_N) = \frac{1}{N!} \sum_{\{\}} f(\{v_1, \dots, v_N\})$$

and $\sum_{\{\}}$ means the sum over all permutations of variables within $\{\}$.

Any solution to Eq. (13.58) satisfies

$$\frac{\partial}{\partial t} \tilde{f} = \left(\widetilde{\overline{\Lambda} f + \overline{\Gamma} f}\right), \tag{13.59}$$

where

$$\left(\widetilde{\overline{\Lambda} f + \overline{\Gamma} f}\right) = \overline{\overline{\Lambda}}\, \widetilde{f} + \overline{\Gamma} \widetilde{f},$$

and

$$\overline{\overline{\Lambda}} f\left(t, v_1, v_2, ..., v_N\right) =$$

$$= \sum_{\substack{1 \le n, m \le N \\ n \ne m}} \left(\int\limits_0^\infty A\left(v_n; w, v_m\right) a\left(w, v_m\right) \times \right.$$

$$\times \quad f\left(t, v_1, \dots, v_{n-1}, w, v_{n+1}, \dots, v_N\right) \mathrm{d}w$$

$$\left. - \quad a\left(v_n, v_m\right) f\left(t, v_1, \dots, v_N\right) \right),$$

This leads to the subsequent equation

$$\frac{\partial}{\partial t} \tilde{f} = \overline{\overline{\Lambda}} \tilde{f} + \overline{\Gamma} \tilde{f}. \tag{13.60}$$

Rescaling the first term on the right–hand side of Eq. (13.60)

$$\frac{\partial}{\partial t} \tilde{f} = \frac{1}{N} \overline{\overline{\Lambda}} \tilde{f} + \overline{\Gamma} \tilde{f} \tag{13.61}$$

we can deal with the limit $N \to \infty$ in a classical way; see Chapter 9 and Refs. [24, 130].

As in Eq. (9.7), we introduce the s–individual marginal density ($1 \le s < N$)

$$\tilde{f}^s\left(t, v_1, \dots, v_s\right) = \left(\int\limits_0^\infty \right)^{N-s} \tilde{f}\left(v_1, \dots, v_N\right) \mathrm{d}v_{s+1} \dots \mathrm{d}v_N,$$

where $\tilde{f}^N = \tilde{f}$, $\tilde{f}^{s'} = 0$ if $s' > N$ and we applied Convention Eq. (2.11).

We assume that the initial data (for $t = 0$) are chaotic, i.e. factorized, probability density

$$\tilde{\mathring{f}} = \underbrace{\mathring{f}^{[1]} \otimes \dots \otimes \mathring{f}^{[1]}}_{N \times}, \tag{13.62}$$

where $\mathring{f}^{[1]} \in L^1$ is a probability density on $[0, \infty)$.

The propagation of chaos is preserved; see Chapter 9, and the solution $f^{\star,1}(t)$ can be linked, in the limit $N \to \infty$, to the solution of the following nonlinear equation

$$\begin{aligned}
\frac{\partial}{\partial t} f^{[1]}(t,v) &= \\
&= \int_0^\infty\int_0^\infty A(v;w,w^*)a(w,w^*)f^{[1]}(t,w)f^{[1]}(t,w^*)\,\mathrm{d}w\,\mathrm{d}w^* \\
&- f^{[1]}(t,v)\int_0^\infty a(v,w)f^{[1]}(t,w)\,\mathrm{d}w + \\
&+ \int_0^\infty B(v;w)b(w)f^{[1]}(t,w)\,\mathrm{d}w - b(v)f^{[1]}(t,v)\,, \qquad v \in [0,\infty)\,.
\end{aligned} \tag{13.63}$$

As in Chapter 9, the following statement holds; cf. Ref. [66]

Theorem 39 *Let $\overset{\circ}{f}{}^{[1]}$ be a probability density on $[0,\infty)$. Then, for each $\tau > 0$, there exists N_0 such that for $N \geq N_0$,*

$$\sup_{t\in[0,\tau]} \|\tilde{f}^1(t) - f^{[1]}(t)\|_1 \leq \frac{\mathfrak{c}}{N^\varsigma}\,,$$

where $\tilde{f}(t) \in L^1_{(N)}$, $t \in [0,\tau]$, is the unique nonnegative solution of Eq. (13.61), corresponding to the chaotic initial data Eq. (13.62), and $f^{[1]}(t) \in L^1$, $t \in [0,\tau]$, is the unique nonnegative solution of Eq. (13.63) corresponding to the initial datum $\overset{\circ}{f}{}^{[1]}$; and ζ, $\mathfrak{c}$ are positive constants that depend on τ.

As discussed in Section 9.3, Theorem 39 demonstrates the connection between the solution of Eq. (13.61), and thereby with the solution of Eq. (13.59), and the macroscopic S–shaped curves commonly seen in experimental observations; see, e.g., Refs. [10, 177, 187, 188, 218].

Let ψ be a given nonnegative bounded regular function on $[0,\infty)$. Assume that; cf. Assumption 12,

Assumption 19

$$\int_0^\infty \psi(v)A(v;w,z)\,\mathrm{d}v = \mathfrak{c}_1\psi(w)\,, \qquad \int_0^\infty \psi(v)B(v;w)\,\mathrm{d}v = \mathfrak{c}_2\psi(w)$$

$$a(v,w) = \psi(w)\,, \qquad b(w) = \psi(w) \qquad \forall\ \ v,w \geq 0\,,$$

where c_1, c_2 are positive constants.

If Assumption 19 holds and $\mathfrak{c}_1 > 1$, $0 < \mathfrak{c}_2 < 1$, the moment of the solution of Eq. (13.63)

$$f^{[1]*}(t) = \int_0^\infty \psi(v)f^{[1]}(t,v)\,\mathrm{d}v$$

satisfies the logistic equation

$$\frac{\mathrm{d}}{\mathrm{d}t}\varrho = (\mathfrak{c}_1 - 1)\,\varrho - (1 - \mathfrak{c}_2)\,\varrho^2\,,$$

and thus results in curves of shape S for reasonably selected initial data.

13.4 Relocation in a closed domain

Following Ref. [76], we consider an equation at the microscopic scale that describes the relocation of agents (persons or other individuals) within a confined space like an elevator. The modelling refers to experiments conducted to explore the dynamics between pedestrians, as detailed in Ref. [87] and other publications cited in Ref. [76].

In various experiments, researchers examined the movement of agents entering a confined space, particularly by investigating the dynamics of evacuation or concentration within such a space. The interested reader is referred to the vast literature cited in Refs. [61, 70, 71, 87, 89, 119, 153, 182]. Extensive research has been carried out on the topic of various aspects of crowd dynamics, with models that usually span various scales, including macroscopic (e.g. Ref. [61, 182]), mesoscopic; see e.g. Ref. [29], microscopic; see e.g. Ref. [119], and even at the nanoscale; see Ref. [61]. New types of experiment have recently been conducted to explore the interactions among pedestrians; see Refs. [87, 88, 153]. Reference [182] deals with macroscopic models of vehicle and pedestrian dynamics, allowing a comprehensive understanding of key characteristics commonly observed in urban traffic.

In Ref. [76] the n–th agent is characterized by its position $x_n \in \mathbb{D}$, $n = 1, \ldots, N$. The model discussed in [76] and its analysis are in the initial stages and still require further development based on experimental data. However, Ref. [76] seems to be the first attempt to formulate a mathematical model through a differential equation that characterizes the behaviour in time of the stochastic relocation situation in the given context. The model is introduced in line of Chapter 9 and provides a description of the microscopic stochastic distribution related to each agent within the system, and the system's state is characterized by the locations of all agents. The model is microscopic, that is, individually based. Two categories of mechanisms are taken into account, one that describes the impact of the domain, which includes the wall and elevator entrance; and the other that describes the interactions among agents, decisive in choosing a position. The former refers to the tendency to concentrate close to the entrance and the tendency to stay close to the boundary. The latter refers to the tendency to maintain a certain distance from other individuals. The latter is connected to the idea of personal space; Ref. [11, 195] and to cultural and social factors; see Ref. [107].

13.4.1 Relocation: Microscopic scale—general model

We apply the general methodology described in Chapter 9 at the microscopic scale to describe the dynamics of individuals entering and moving within a confined space like an elevator. We take into account a system of N ($N < \infty$) agents, where the agent n–th, with $n = 1, \ldots, N$, is identified by its location

$$x_n \in \mathbb{D}\,,$$

outside $\mathbb{D}_{\text{out}}$ or inside $\mathbb{D}_{\text{in}}$ the elevator,

$$\mathbb{D}_{\text{out}} \cup \mathbb{D}_{\text{in}} = \mathbb{D}\,, \qquad \mathbb{D}_{\text{out}} \cap \mathbb{D}_{\text{in}} = \emptyset\,.$$

The evolution of the system is given by the changes over time of the probability density related to the positions of all N interacting agents. The evolution equation is defined by a linear generator; like in Eq. (9.3), that reflects the interactions and fully determines the behaviour of the system on the microscopic scale for any $t \geq 0$.

The transition of the position of the n_1–agent occurs at a random time, and there are the following possibilities.

- The transition happens independently without any interaction with other agents: any agent either enters and selects its position or exits.
- The transition occurs due to the interaction with the agents n_2, n_3, ..., n_N

$$n_2, \ldots, n_N \in \{1, \ldots, N\}\,.$$

The rate of interaction that causes the change of the position of agent n_1 with the position x_{n_1} entering the elevator and choosing the position or exiting is given by a function $a^{[1]} = a^{[1]}(x_{n_1})$. The choice of position is described by $A^{[1]} = A^{[1]}(x; x_{n_1})$.

We assume

Assumption 20 *Let*

- $a^{[1]}$ *be a measurable function such that*

$$0 \leq a^{[1]}(x_{n_1}) \leq \mathfrak{c} < \infty\,, \qquad \forall\, x_{n_1} \in \mathbb{D}\,, \tag{13.64}$$

where $\mathfrak{c} > 0$ *is a constant.*

- $A^{[1]}$ *be a measurable function such that*

$$A^{[1]} = A^{[1]}(x; x_{n_1}) \geq 0\,, \qquad \forall\, x, x_{n_1} \in \mathbb{D}\,, \tag{13.65}$$

and

$$\int_{\mathbb{D}} A^{[1]}(x\,;\, x_{n_1})\, \mathrm{d}x = 1\,, \tag{13.66}$$

for all $x_{n_1} \in \mathbb{D}$ *such that*

$$a^{[1]}(\,x_{n_1}\,) > 0\,.$$

Then, the rate of interaction between the agent n_1 with position x_{n_1} and the agents n_2, ..., n_N with positions x_{n_2}, ..., x_{n_N}, is given by the function $a^{[N]} = a^{[N]}(x_1, \ldots, x_N)$. The change of position into the point x of the agent n_1 with initial position x_{n_1}, due to the interaction with agents n_2, ..., n_N with positions x_{n_2}, ..., x_{n_N}, respectively, is described by a measurable function $A^{[N]} = A^{[N]}(x; x_{n_1}, \ldots, x_{n_N})$.

We assume

Assumption 21 *Let*

- $a^{[N]}$ *is a measurable function such that*

$$0 \leq a^{[N]}(x_{n_1}, x_{n_2}, \dots, x_{n_N}) \leq \mathfrak{c} < \infty, \qquad \forall\, x_{n_1}, \dots, x_{n_N} \in \mathbb{D}, \tag{13.67}$$

where $\mathfrak{c} > 0$ *is a constant.*

- $A^{[N]}$ *is a measurable function such that*

$$A^{[N]} = A^{[N]}(x; x_{n_1}, \dots, x_{n_N}) \geq 0, \qquad \forall\, x, x_{n_1}, \dots, x_{n_N} \in \mathbb{D}, \tag{13.68}$$

and

$$\int\limits_{\mathbb{D}} A^{[N]}(x\,; x_{n_1}, x_{n_2}, \dots, x_{n_N})\, \mathrm{d}x = 1, \tag{13.69}$$

for all $x_{n_1}, \dots, x_{n_N} \in \mathbb{D}$ *such that*

$$a^{[N]}(x_{n_1}, \dots, x_{n_N}) > 0.$$

Similarly as in Chapter 9 the functions $a^{[1]}$, $a^{[N]}$, $A^{[1]}$, and $a^{[N]}$ completely determine the model at the microscopic level.

Given N, $a^{[1]}$, $a^{[N]}$, $A^{[1]}$, and $a^{[N]}$, following Chapter 9, we consider the stochastic system that is given as the Markov jump process of N agents through the generator Λ

$$\frac{\partial}{\partial t} f = \Lambda f\,; \qquad f\Big|_{t=0} = \overset{\circ}{f}, \tag{13.70}$$

where Λ has the form

$$\begin{aligned}
&\Lambda f(t, x_1, x_2, ..., x_N) = \\
&= \sum_{1 \leq n \leq N} \left(\int\limits_{\mathbb{D}} A^{[1]}(x_n; x) a^{[1]}(x) f(t, x_1, \dots, x_{n-1}, x, x_{n+1}, \dots, x_N)\, \mathrm{d}x \right. \\
&- \left. a^{[1]}(x_n) f(t, x_1, \dots, x_N) \right) + \\
&+ \sum_{\substack{1 \leq n_1, \dots, n_N \leq N \\ n_i \neq n_j\ \forall i \neq j}} \left(\int\limits_{\mathbb{D}} A^{[N]}(x_{n_1}; x, x_{n_2}, \dots, x_{n_N}) a^{[N]}(x, x_{n_2}, \dots, x_{n_N}) \times \right. \\
&\times\ f(t, x_1, \dots, x_{n_1 - 1}, x, x_{n_1 + 1}, \dots, x_N)\, \mathrm{d}x \\
&- \left. a^{[N]}(x_{n_1}, x_{n_2}, \dots, x_{n_N}) f(t, x_1, \dots, x_N) \right).
\end{aligned}$$

As in Chapter 9, Assumptions Eqs. (13.67), (13.68), (13.69) guarantee that the operator Λ is a bounded linear operator in space $L^1_{(N)}$. Therefore, the Cauchy problem Eq. (13.70) together with the initial datum $\overset{\circ}{f}$ has a unique solution in $L^1_{(N)}$ for all $t \geq 0$ given by a uniformly continuous semigroup according to the formula

$$f(t) = e^{t\Lambda}\, \overset{\circ}{f}.$$

The semigroup is a uniformly continuous stochastic semigroup.

13.4.2 Relocation: Mesoscopic scale

Although the number of agents N in the case of an elevator cannot be very large, we can consider the corresponding mesoscopic model as an interesting alternative. The linear equation (13.70) may result; see Chapter 9, in a nonlinear integro–differential equation corresponding to the mesoscopic description,

$$\frac{\partial}{\partial t} f(t,x) = \mathcal{G}[f](t,x) - f(t,x)\mathcal{L}[f](t,x)\,, \qquad x \in \mathbb{D}\,, \tag{13.71}$$

where $\mathcal{G}[f]$ is the gain term, given by

$$\begin{aligned}\mathcal{G}[f](t,x) = &\int_{\mathbb{D}} A^{[1]}(x;y)a^{[1]}(y)f(t,y)\,\mathrm{d}y + \\ + &\sum_{\{\}}\int_{\mathbb{D}^N} A^{[N]}(x;x_1,\{x_2,\ldots,x_N\})a^{[N]}(x_1,\{x_2,\ldots,x_N\}) \times \\ \times &\; f(t,x_1)f(t,x_2)\ldots f(t,x_N)\,\mathrm{d}x_1\,\mathrm{d}x_2\ldots\mathrm{d}x_N\,,\end{aligned}$$

and $f\,\mathcal{L}[f]$ is the loss term, defined as

$$\begin{aligned}\mathcal{L}[f](t,x) = &\; a^{[1]}(x) + \\ + &\sum_{\{\}}\int_{\mathbb{D}^{N-1}} a^{[N]}(x,\{x_2,\ldots x_N\})f(t,x_2)\ldots f(t,x_N)\,\mathrm{d}x_2\ldots\mathrm{d}x_N\,,\end{aligned}$$

where $\sum_{\{\}}$ means the sum over all permutation of variables within $\{\}$.

13.4.3 Relocation: Model

Referring to experimental observations regarding people entering an elevator and relocating themselves within, see Refs. [87–89, 153], it is possible to establish, as previously indicated, fundamental hypotheses: when entering enclosed spaces, people tend to prefer to stand near the walls, but also not too close to the entrance area. However, simultaneously, they aim to reduce the cost of exiting by situating themselves not far from the entrance. It is crucial to take into account the psychological effects that have emerged from the experiments in order to formulate appropriate assumptions regarding the functions defining the model, that is, $a^{[1]}$, $a^{[N]}$, $A^{[1]}$, and $a^{[N]}$ in Eq. (13.70). These considerations should be complemented by more detailed information, possibly from targeted experiments. In the preliminary analysis, we can consider that the total number of people N inside the elevator remains unchanged, thus ignoring any entrances or exits; in this case, it is $\mathbb{D}_{\text{out}} = \emptyset$. Thus, $\mathbb{D} = \mathbb{D}_{\text{in}}$. Additionally, we may consider the subsequent simplifying assumptions

$$a^{[1]} = \text{const} > 0\,, \tag{13.72}$$

and

$$a^{[N]} = \text{const} > 0\,. \tag{13.73}$$

Let the probability density $A^{[1]}(\,.\,,x_1)$ be independent of x_1. It expresses the probability of taking the position without interactions with other individuals. We assume that

$$A^{[1]}(x) = \lambda_1 A_1^{[1]}(x) + \lambda_2 A_2^{[1]}(x)\,, \qquad \lambda_i \geq 0 \quad \text{and} \quad \lambda_1 + \lambda_2 = 1\,, \tag{13.74}$$

where probability densities on $\mathbb{D}$, $A_1^{[1]}$, and $A_2^{[1]}$, represent the tendency to concentrate close to the entrance and the tendency to be close to the boundary, respectively.

The function $A^{[N]}$ describes the interactions between agents in the sense of a tendency not to be too close to another agent. The function $A^{[N]} = A^{[N]}(x; x_1, \ldots, x_N)$ is a probability density with respect to x variable (for all $x_1, x_2, \ldots, x_N$ in $\mathbb{D}$) on $\mathbb{D}$ such that

$$A^{[N]}(x; x_1, x_2 \ldots, x_N) = 0 \quad \text{on} \quad \mathbb{D}_{x_2, \ldots, x_N}, \tag{13.75}$$

where $\mathbb{D}_{x_2, \ldots, x_N} = \mathbb{D} \cap \bigcup_{i=2,\ldots,N} \mathbb{B}_i$, and $\mathbb{B}_i$ is a ball with the centre at x_i and the small radius $\delta > 0$ such that $2N\pi\delta^2 < |\mathbb{D}|$. Moreover, we assume that $A^{[N]} = A^{[N]}(x; x_1, \ldots, x_N)$ is a quickly decreasing function of $|x - x_1|$.

The specific assumptions on the quantities $A^{[1]}(x)$ and $A^{[N]}(x; x_1, \ldots, x_N)$ lead to the model of relocation of individuals inside the "elevator". This model is definitely very preliminary and should be refined taking into account information from experiments.

Following Chapter 9, we obtain the qualitative analysis of the model; see Ref. [76].

Assumption 22 *Let at least one of the following two conditions be satisfied.*

$$A^{[1]}(x_n; x) a^{[1]}(x) \geq \mathfrak{c} > 0 \tag{13.76}$$

for all $x, x_n \in \mathbb{D}$;

or

$$A^{[N]}(x_{n_1}; x, x_{n_2}, \ldots, x_{n_N}) a^{[N]}(x, x_{n_2}, \ldots, x_{n_m}) \geq \mathfrak{c} > 0 \tag{13.77}$$

for all $x_{n_1}, x_{n_2}, x \in \mathbb{D}$, where $\mathfrak{c} > 0$ is a constant.

It is evident that Assumption (13.76) is reasonable in the given situation, while Assumption (13.77) appears to be quite restrictive. However, in a simplified situation, where $\mathbb{D}_{\text{out}} = \emptyset$ it may also be natural to some extent.

We can see that $\left(e^{t\Lambda}\right)_{t\geq 0}$ is partially integral and under Assumption 22 we have; see Ref. [76],

Theorem 40 *Let*

- *Assumptions given in Eqs.* (13.64) *and* (13.69) *be satisfied,*
- *Assumption 22 be satisfied.*

Then the semigroup $\left(e^{t\Lambda}\right)_{t\geq 0}$ is asymptotically stable.

Proof. See Section 9.2. ■ QED

Research on the inflow process is not as developed as that on the outflow process, despite the potential close connection between the two. Conducting a thorough investigation of the inflow process may offer a fuller depiction of scenarios pertaining to emergencies, such as evacuation procedures or rules of conduct in the event of an earthquake. We presented here a very preliminary stochastic microscopic model and a related model at the mesoscopic level. It is hoped that a more accurate model may provide important information about the process under study.

Chapter 14

Conclusion

Many processes in the life and social sciences require a precise description of potential interactions between individual agents of the given system rather than a description of the system as a whole. Typical examples given in Chapters 1 and 2 are creations of swarms, flocks, herds, etc., cf. Refs. [17, 21, 38, 42, 155]. Interactions result in jumps of the states that characterize agents. In many situations, the interactions demonstrate a clear random character. In this monograph we presented a basis for a mathematical general framework that can be used to model behaviours of specific processes in the life and social sciences. The behaviour of the solutions was quite complex and could imitate the behaviour of various real systems. We can observe either a trend towards order or disorder—see Proposition 2. Another effect that may characterize such processes is the non–local nature of the interactions—cf. Chapter 12. I believe that the mathematical theory presented in this monograph could be a starting point to provide a better mathematical description of life and social phenomena in which interactions between agents are essential.

This monograph presents many issues that require further mathematical research. For example, for $\gamma \in (0, 1)$, both in the continuous and discrete cases, only very basic results are known—cf. Chapter 3. This case, apart from its interesting mathematical structure, may be suitable for describing important processes in life and social sciences. Chapter 4 presents the complete theory of the simplified case of a constant interaction kernel. In the more realistic and more general case of a variable kernel, the results obtained are only partial; see Chapter 5. Similarly, the results of Chapter 6 require further generalizations and the development of new mathematical methods. In particular, the emergence of bimodal distributions requires further research motivated by specific applications. The relationships with the partial differential equations; cf. Chapter 7, are an important topic for further research. Such research may be related to the non–local models described in Section 12.3. There are numerous open problems in the study of the relationship between integro–differential equations and the corresponding partial differential equations. An example could be Conjecture 1 below, stated in Ref. [134] and related to the methods in Chapter 7.

Conjecture 1 *Assume*

- $\mathfrak{n} = \gamma = \nu = 1$,
- $\overset{\circ}{f}$ *is a probability density in* $\mathbb{V}$ *such that* $\overset{\circ}{f} \in X^{(3)}$*;*
- $a \in X^{(3)}$ *satisfies Assumption* (8) *and*

$$\int_{\mathbb{V}} a(v)|v|^3 \,\mathrm{d}v \leq \mathfrak{c},$$

  for some positive constant $\mathfrak{c}$*;*
- $f^{(\varepsilon)} = f^{(\varepsilon)}(t)$*, defined on* $[0, \tau]$*, is a solution of Eq.* (7.19) *with the initial datum* $\overset{\circ}{f}$*;*

DOI: 10.1201/9781003537595-14

- *$f_B = f_B(t)$, defined on $[0,\tau]$, in $X^{(1)}$ is a solution to Eq. (7.22) with the same initial datum $\overset{\circ}{f}$.*

Then

$$\sup_{t\in[0,\tau]} \left| \int_{\mathbb{V}} \varphi(v) \left(f^{(\varepsilon)}(t,v) - f_B(t,v) \right) \mathrm{d}v \right| \leq \mathfrak{c}\,\varepsilon ,$$

for each $\varphi \in C^{(3)}(\mathbb{V})$, where the constant $\mathfrak{c}$ can depend on φ but does not depend on ε.

Extended research from Chapter 8 may be the basis for developing the theory of macroscopic limits, hydrodynamic limits; cf. for example B. Perthame's book [175]. In the context of the equations considered in this monograph, the first step was taken in Ref. [170].

It is possible that future studies will encompass even more general mathematical structures than those considered in the present monograph. For example we can consider the following general mesoscopic equation

$$\frac{\partial}{\partial t} f(t,x,v) = \int_{\mathbb{V}}\int_{\mathbb{D}} \Big(\mathcal{T}_f(y,v;x,w) - \mathcal{T}_f(y,w;x,v) \Big)\, \mathrm{d}y\ \mathrm{d}w , \tag{14.1}$$

where $T_f(y,v;x,w)$, as before, describes the transition from state $w \in \mathbb{V}$ to state $v \in \mathbb{V}$ of an agent at place $x \in \mathbb{D}$ due to interaction with an agent at $y \in \mathbb{D}$ and a choice of the function $\mathcal{T}_f$ can depend in a general way on f

$$\mathcal{T}_f(y,v;x,w) = a(y,v;x,w)\, \mathcal{F}[f(t,y,v)]\, \mathcal{G}[f(t,x,w)] ,$$

where $\mathcal{F}$, $\mathcal{G}$: $\mathbb{R}_+ \mapsto \mathbb{R}_+$ are suitable chosen sufficiently regular functions.

The previous natural assumption was

$$\mathcal{F}[x] = x^{\gamma} , \qquad \mathcal{G}[x] = x , \qquad x \geq 0 .$$

On the other hand, the general class of equations now reads

$$\begin{aligned} \frac{\partial}{\partial t} f(t,x,v) &= \\ &= \int_{\mathbb{D}}\int_{\mathbb{V}} a(y,v;x,w)\, \mathcal{F}[f(t,y,v)]\, \mathcal{G}[f(t,x,w)]\, \mathrm{d}y\ \mathrm{d}w \\ &- \mathcal{G}[f(t,x,v)] \int_{\mathbb{D}}\int_{\mathbb{V}} a(y,w;x,v)\, \mathcal{F}[f(t,y,w)]\, \mathrm{d}y\ \mathrm{d}w . \end{aligned} \tag{14.2}$$

The function $a(y,v,x,w)$ is again the rate (kernel) of interaction between an agent with state w at the point x and an agent with state v located at the point y and describes the transition from orientation w to orientation v.

Independently we can propose a stochastic individually based (microscopic) model following the idea stated in Chapter 9 where the system of N agents is taken into account in the modelling process. Such a system is described in terms of a Markov jump process and related linear evolution equations. The equation describes the evolution of probability densities, with a microscopic representation of the system of N interacting agents.

Then we can provide a mathematical approach that links in the limit $N \to \infty$, the solutions of the linear equation with the solutions of the corresponding general nonlinear integro–differential equation that can be related to the mesoscopic description defined by the general mesoscopic equation (14.1).

The possible rich behaviour of solutions of Eq. (13.10) given in Chapter 4 can lead to blow–ups in a finite time. This behaviour is not possible for the mesoscopic equations that result directly from the microscopic structures in Chapter 9. This may lead to interesting mathematical questions on the behaviour of solutions when approximating delta distributions.

The case of discrete equations discussed in Chapter 11 clearly requires further research, especially since it is related to many equations already found in biomathematics and game theory. The difficulties that will arise here are typical for systems of ordinary equations in higher dimensions. Therefore, the results may have great implications for the development of finite–dimensional dynamic systems as well as for modelling in the life and social sciences.

The most interesting generalizations may come from considering the various possible applications of the equations discussed. The mathematical structures considered in the monograph are based on the description of interactions between agents of a given population (or various populations). As a result, descriptions of this type are common and can be used for any process when changes in the individual states of agents occur as a result of the interaction between them. New applications will inspire new mathematical research and new results will provide opportunities for a better description of processes in the life and social sciences.

Concluding proverb

Sicilian proverb

Ciò ch'é fattu, e 'un si nni parra cchiú.

Bibliography

[1] V.I. Abkevich, A.M. Gutin, and E.I. Shakhnovich. Impact of local and non–local interactions on thermodynamics and kinetics of protein folding. *J. Molecular Biol.*, 252(4):460–471, 1995.

[2] G. Abramson, C.A. Trejo Soto, and L. Oña. The role of asymmetric interactions on the effect of habitat destruction in mutualistic networks. *PLoS ONE*, 6(6), 2011.

[3] J.A. Adam and N. Bellomo, editors. *A Survey of Models for Tumor-Immune System Dynamics.* Birkhäuser, Boston, 1996.

[4] R.A. Adams. *Sobolev Spaces.* Academic Press, New York, 1975.

[5] O.E. Adebayo, D. Trucu, and R. Eftimie. Analytical investigation of a non–local mathematical model for normal and abnormal wound healing. *Discrete Cont. Dynam. Systems – B*, pages 1–18, 2024.

[6] G. Ajmone-Marsan, N. Bellomo, and M. Egidi. Towards a mathematical theory of complex socio–economical systems by functional subsystems representation. *Kinetic Rel. Models*, 1(2):249–278, 2008.

[7] G. Ajmone-Marsan, N. Bellomo, and L. Gibelli. Stochastic evolutionary differential games toward a system theory of behavioral social dynamics. *Math. Models Methods Appl. Sci.*, 26(06):1051–1093, 2016.

[8] G. Akerlof. The market for "lemons": Quality uncertainty and the market mechanism. *Quarterly J. Econ.*, 84(3):488–500, 1970.

[9] G. Albi and L. Pareschi. Modelling self–organized systems interacting with few individuals: from microscopic to macroscopic dynamics. *Applied Math. Letters*, 26:397–401, 2013.

[10] A.E. Allahverdyan, Z.H.S. Gevorkian, C.-K. Hu, and Th.M. Nieuwenhuizeni. How absorption influences DNA denaturation. *Phys. Rev. E*, 79:4988–4991, 2000.

[11] I. Altman. *The environment and social behavior: privacy, personal space, territory, crowding.* Brooks–Cole Pub. Co., 1975.

[12] F. Andreu-Vaillo, J. Mazon, J. Rossi, and J.J Toledo-Melero. *Nonlocal Diffusion Problems*, volume 165 of *Mathematical Surveys and Monographs.* American Mathematical Society, Providence, Rhode Island, 2010.

[13] L. Arlotti, N. Bellomo, E. De Angelis, and M. Lachowicz. *Generalized Kinetic Models in Applied Sciences.* World Sci., Singapore, 2003.

[14] L. Arlotti, A. Gamba, and M. Lachowicz. A kinetic model of tumor immune system cellular interactions. *J. Theor. Medicine*, 4:39–50, 2002.

[15] D.G. Aronson. *The porous medium equation*, volume 1224, chapter Nonlinear Diffusion Problems, pages 1–46. Lecture Notes in Math., Springer, 1986.

[16] K.J. Arrow. A difficulty in the concept of social welfare. *J. Political Economy*, 58:328–346, 1950.

[17] A. Attanasi, A. Cavagna, L. Del Castello, I. Giardina, T.S. Grigera, A. Jelić, S. Melillo, L. Parisi, O. Pohl, and E. Shen M.J. Viale. Information transfer and behavioural inertia in starling flocks. *Nat Phys.*, 1;10(9):615–698, 2014.

[18] B. Aylaj, N. Bellomo, L. Gibelli, and A. Reali. A unified multiscale vision of behavioral crowds. *Math. Models Methods Appl. Sci.*, 30:1–22, 2020.

[19] S. Baigent. Lotka–volterra dynamics: An introduction. *Semantic Scholar*, 2016.

[20] M.S. Balbuena, G.T. Broadhead, A. Dahake, E. Barnett, M. Vergara, K.A. Skogen, T. Jogesh, and R.A. Raguso. Mutualism has its limits: consequences of symmetric interactions between a well–defended plant and its herbivorous pollinator. *Phil. Trans. R. Soc. B*, 377(1853), 2022.

[21] M. Ballerini, N. Cabibbo, R. Candelier, A. Cavagna, E. Cisbani, I. Giardina, V. Lecomte, A. Orlandi, G. Parisi, A. Procaccini, M. Viale, and V. Zdravkovic. Interaction ruling animal collective behavior depends on topological rather than metric distance: Evidence from a field study. *Proc. Nat. Acad. Sci.*, 105(4):1232–1237, 208.

[22] J. Banasiak. Kinetic models – mathematical models of everything? *Phys. Life Rev.*, 16, 01 2016.

[23] J. Banasiak and M. Lachowicz. On a macroscopic limit of a kinetic model of alignment. *Math. Models Methods Appl. Sci.*, 23(14):2647–2670, 2013.

[24] J. Banasiak and M. Lachowicz. *Methods of small parameter in mathematical biology.* Birkhäuser Cham, Boston, 2014.

[25] M. Banerjee, S. Ghosh, P. Manfredi, and A. d'Onofrio. Spatio–temporal chaos and clustering induced by nonlocal information and vaccine hesitancy in the sir epidemic model. *Chaos, Solitons Fractals*, 170:113339, 2023.

[26] V. Barbu. *Nonlinear Semigroups and Differential Equations in Banach Spaces.* Noordhoff, Leyden, IL, USA, 1976.

[27] A. Bejan and S. Lorente. The constructal law origin of the logistics s curve. *J. Appl. Phys.*, 110:024901, 2011.

[28] N. Bellomo, G. Ajmone Marsan, and A. Tosin. *Complex Systems and Society. Modeling and Simulation.* Springer, New York, 2013.

[29] N. Bellomo, A. Bellouquid, and D. Knopoff. From the microscale to collective crowd dynamics. *Multiscale Model. Simul.*, 11:943–963, 2013.

[30] N. Bellomo, A. Bellouquid, Y. Tao, and M. Winkler. Toward a mathematical theory of Keller-Segel models of pattern formation in biological tissues. *Math. Models Methods Appl. Sci.*, 25(9):1663–1763, 2015.

[31] N. Bellomo, D. Burini, V. Secchini, and P. Terna. *Active Particles Methods in Economics: New Perspectives in the Interaction between Mathematics and Economics.* Elements in Complexity and Agent-based Economics. Cambridge University Press, 2025.

[32] N. Bellomo and B. Carbonaro. Toward a mathematical theory of living systems focusing on developmental biology and evolution: A review and perspectives. *Phys. Life Rev.*, 8:1–18, 2011.

[33] N. Bellomo, P. Degond, and E. Tadmor, editors. *Active Particles, Volume 1. Advances in Theory, Models, and Applications.* Birkhäuser Cham, 2017.

[34] N. Bellomo, L. Gibelli, and N. Outada. On the interplay between behavioral dynamics and social interaction in human crowds. *Kinet. Relat. Models*, 12:397–409, 2019.

[35] N. Bellomo, N.K. Li, and P. Maini. On the foundations of cancer modeling: selected topics, speculations and perspectives. *Math. Models Methods Appl. Sci.*, 18, 2008.

[36] N. Bellomo, N. Outada, J. Soler, Y. Tao., and M. Winkler. Chemotaxis and cross–diffusion models in complex environments: models and analytic problems toward a multiscale vision. *Math. Models Methods Appl. Sci.*, 32(4):713–792, 2022.

[37] N. Bellomo and M. Pulvirenti, editors. *Modeling in Applied Sciences. Kinetic Theory Approach.* Birkhäuser, Boston, 2013.

[38] D. Benedetto, T. Paul, and S. Rossi. Propagation of chaos and hydrodynamic description for topological models. *Kinet. Relat. Models*, 18(1):19–34, 2025.

[39] Ph. Benilan, M.G. Crandall, and M. Pierre. Solutions of porous medium equation in R^N under optimal conditions on initial values. *Indiana Univ. Math. J.*, 33(1):51–87, 1984.

[40] M. Bisi, M. Groppi, G. Martalò, and R. Travaglini. Derivation from kinetic theory and 2-d pattern analysis of chemotaxis models for multiple sclerosis. *Arxiv*, arXiv:2501.13119v1 [q-bio.QM] 20 Jan 2025, 2025.

[41] M. Bisi and R. Travaglini. Reaction–diffusion equations derived from kinetic models and their turing instability. *Communic. Math. Sci.*, 20:763–801, 2022.

[42] A. Blanchet and P. Degond. Topological interactions in a Boltzmann–type framework. *J. Stat. Phys.*, 163:41–60, 2016.

[43] M. Bodnar, M.J. Piotrowska, and U. Foryś. Existence and stability of oscillating solutions for a class of delay differential equations. *Nonlinear Anal.: Real World Applications*, 14(3):1780–1794, 2013.

[44] F. Bouchut, F. Golse, and M. Pulvirenti. *Kinetic equations and asymptotic theory.* Series in Applied Mathematics. Elsevier, Editors: B. Perthame and L. Desvillettes, 2000.

[45] F. Brauer and C. Castillo–Chavez. *Mathematical Models in Population Biology and Epidemiology.* Springer, New York, 2012.

[46] M. Braun. *Differential Equations and Their Applications.* Springer, New York, 1975.

[47] D. Burini, E. De Angelis, and M. Lachowicz. A continuous–time markov chain modeling cancer–immune system interactions. *Commun. Appl. Ind. Math.*, 9(2):106–118, 2018.

[48] E. Carlen, P. Degond, and B. Wennberg. Kinetic limits for pair–interaction driven master equation and biological swarm models. *Math. Models Methods Appl. Sci.*, pages 1339–1376, 2013.

[49] J. A. Carrillo, M. Fornasier, G. Toscani, and F. Vecil. *Particle, Kinetic, and Hydrodynamic Models of Swarming*, chapter Mathematical Modeling of Collective Behavior in Socio–Economic and Life Sciences, pages 297–336. Birkhäuser, Boston, 2010.

[50] J. A. Carrillo, M. Parisot, and Z. Szymańska. Mathematical modelling of collagen fibres rearrangement during the tendon healing process. *Kinet. Relat. Models*, 14(2):283–301, 2021.

[51] J.A. Carrillo, M.R. D'Orsogna, and V. Panferov. Double milling in self–propelled swarms from kinetic theory. *Kinet. Relat. Models*, 2:363–378, 2009.

[52] J.A. Carrillo, A. Esposito, and J. Sh. Wu. Nonlocal approximation of nonlinear diffusion equations. *Calc. Var.*, 63(100), 2024.

[53] J.A. Carrillo and E. Tadmor. *Active Particles, Volume 4*. Birkhäuser, Cham, 2024.

[54] C. Cercignani, R. Illner, and M. Pulvirenti. *The Mathematical Theory of Dilute Gases*. Springer, New York, 1994.

[55] M. Chaplain, M. Lachowicz, Z. Szymańska, and D. Wrzosek. Mathematical modelling of cancer invasion: The importance of cell–cell adhesion and cell–matrix adhesion. *Models Methods Appl. Sci.*, 24(4):719–743, 2011.

[56] A. Chauviere and I. Brazzoli. On the discrete kinetic theory for active particles. mathematical tools. *Math. Computer Model.*, 43(7):933–944, 2006.

[57] A. Ching and S. Baigent. The balance simplex in non–competitive 2–species scaled lotka–volterra systems. *J. Biol. Dyn.*, 13(1):128–147, 2019.

[58] C. Christopher and J. Devlin. A note on the replicator equations of dynamical game theory. *Appl. Math. Lett.*, 6(4):31–33, 1993.

[59] S. Cordier, L. Pareschi, and G. Toscani. On a kinetic model for a simple market economy. *J. Statist. Phys.*, 120, 2005.

[60] M.F.P. Costa, M.P. Ramos, C. Ribeiro, and A.J. Soares. Optimal control model of immunotherapy for autoimmune diseases. *Math. Methods Appl. Sci.*, 44(11):8883–8902, 2021.

[61] E. Cristiani, B. Piccoli, and A. Tosin. *Multiscale Modeling of Pedestrian Dynamics*. Springer, 2014.

[62] E. Cristiani, F.S. Priuli, and A. Tosin. Modeling rationality to control self-organization of crowds: an environmental approach. *SIAM J. Appl. Math.*, 75(2):605–629, 2015.

[63] N. David and B. Perthame. Free boundary limit of tumor growth model with nutrient. *J. Math. Pures Appl.*, 155:62–82, 2021.

[64] M. Dębowski. Diauxic behaviour for biological processes at various timescales. *Math. Models Appl. Sci.*, pages 1–9, 2020.

[65] M. Dębowski. *Some aspects of molecular mechanisms of the cell cycle and diauxic growth from a mathematical perspective (In Polish)*. PhD thesis, University of Warsaw, 2024.

[66] M. Dębowski, M. Lachowicz, and Z. Szymańska. Microscopic description of DNA thermal denaturation. *Appl. Math. Computation*, 361:47–60, 2019.

[67] M. Dębowski, Z. Szymańska, J.Z. Kubiak, and M. Lachowicz. Mathematical model explaining the role of cdc6 in the diauxic growth of cdk1 activity during the m-phase of the cell cycle. *Cells*, 8, 2019.

[68] P. Degond, A. Frouvelle, and G. Raoul. Local stability of perfect alignment for a spatially homogeneous kinetic model. *J. Stat. Phys.*, 157:84–112, 2014.

[69] R. Della Marca, M.P. Machado Ramos, C. Ribeiro, and A.J. Soares. Mathematical modelling of oscillating patterns for chronic autoimmune diseases. *Math. Methods Appl. Sci.*, 45(11):7144–7161, 2022.

[70] M. Di Francesco, S. Fagioli, M.D. Rosini, and G. Russo. Deterministic particle approximation of the hughes model in one space dimension. *Kinet. Relat. Models*, 10:215–237, 2017.

[71] M. Di Francesco and M.D. Rosini. Rigorous derivation of nonlinear scalar conservation laws from follow– the–leader type models via many particle limit. *Archive Rational Mech. Anal.*, 217:831–871, 2015.

[72] R.J. DiPerna and P.-L. Lions. On the Cauchy problem for Boltzmann equations: global existence and weak stability. *Ann. Math.*, 2.130:321–366, 1989.

[73] M. Dolfin and M. Lachowicz. Modeling altruism and selfishness in welfare dynamics: The role of nonlinear interactions. *Math. Models Methods Appl. Sci.*, 24(12):2361–2381, 2014.

[74] M. Dolfin and M. Lachowicz. Modeling DNA thermal denaturation at the mesoscopic level. *Discrete Contin. Dyn. Syst. Ser. B*, 19(8):2469–2482, 2014.

[75] M. Dolfin and M. Lachowicz. Modeling opinion dynamics: How the network enhances consensus. *Networks Heterogeneous Media*, 10(4):877–896, 2015.

[76] M. Dolfin, M. Lachowicz, and A. Schadschneider. *A microscopic model of redistribution of individuals inside an "Elevator"*, chapter Modern Problems in Applied Analysis, pages 77–86. Springer, 2018.

[77] A. Dongre and R.A. Weinberg. New insights into the mechanisms of epithelial–mesenchymal transition and implications for cancer. *Nat. Rev. Mol. Cell Biol.*, 20(2):69–84, 2019.

[78] A. d'Onofrio and A. Gandolfi, editors. *Mathematical Oncology 2013*. Birkhäuser, New York, 2014.

[79] A. d'Onofrio, P. Manfredi, and Poletti P. The impact of vaccine side effects on the natural history of immunization programmes: an imitation–game approach. *J. Theor. Biol.*, 21(273(1)):63–71, 2011.

[80] A. d'Onofrio, P. Manfredi, and E. Salinelli. Vaccinating behaviour, information, and the dynamics of sir vaccine preventable diseases. *Theor. Popul. Biol.*, 71(3):301–317, 2007.

[81] Q. Du. *Nonlocal Modeling, Analysis, and Computation*. SIAM, Philadelphia, 2019.

[82] Q. Du, M. Gunzburger, R.B. Lehoucq, and K. Zhou. A nonlocal vector calculus, nonlocal volume-constrained problems, and nonlocal balance laws. *Math. Models Methods Appl. Sci.*, 23(03):493–540, 2013.

[83] G. Dudziuk, M. Lachowicz, H. Leszczyński, and Z. Szymańska. A simple model of collagen remodeling. *Discrete Contin. Dyn. Syst. Ser. B*, 24(5):2205–2217, 2019.

[84] R. Durrett and S. Levin. The importance of being discrete (and spatial). *Theoretical Popul. Biol.*, 46:363–394, 1994.

[85] M. Eckardt, K.J. Painter, Ch. Surulescu, and A. Zhigun. Nonlocal and local models for taxis in cell migration: A rigorous limit procedure. *J. Math. Biology*, 81:1251–1298, 2020.

[86] M. Eckardt and Ch. Surulescu. On a mathematical model for cancer invasion with repellent ph–taxis and nonlocal intraspecific interaction. *arXiv:2307.15342v1 [math.AP] 28 Jul 2023*, 2023.

[87] T. Ezaki, K. Ohtsuka, M. Chraibi, M. Boltes, D. Yanagisawa, A. Seyfried, A. Schadschneider, and K. Nishinari. Inflow process of pedestrians to a confined space. *Collective Dynamics*, 1(A:4):1–18, 2016.

[88] T. Ezaki, K. Ohtsuka, D. Yanagisawa, and K. Nishinari. *Inflow process: A counterpart of evacuation*, chapter Traffic and Granular Flow, pages 227–231. Springer, 2013.

[89] T. Ezaki, D. Yanagisawa, K. Ohtsuka, and K. Nishinari. Simulation of space acquisition process of pedestrians using proxemic floor field model. *Physica A*, 391:291–299, 2012.

[90] M. Fornasier, J. Haskovec, and G. Toscani. Fluid dynamic description of flocking via the Povzner–Boltzmann equation. *Physica D: Nonlinear Phenomena*, 240(1):21–31, 2011.

[91] A. Francou and K.V. Anderson. The epithelial–to–mesenchymal transition in development and cancer. *Ann. Rev. Cancer Biology*, 4:197–220, 2020.

[92] L.C. Franssen and M.A.J. Chaplain. A mathematical multi–organ model for bidirectional epithelial–mesenchymal transitions in the metastatic spread of cancer. *IMA J. Appl. Math.*, 85:724–761, 2020.

[93] L.C. Franssen, T. Lorenzi, A.E.F. Burgess, and M.A.J. Chaplain. A mathematical framework for modelling the metastatic spread of cancer. *Bull. Math. Biol.*, 81:1965–2010, 2019.

[94] J. Furter and M. Grinfeld. Local vs. non–local interactions in population dynamics. *J. Math. Biology*, 27:65–80, 1989.

[95] W.V. Gehrlein. Condorcet's paradox and the likelihood of its occurrence: different perspectives on balanced preferences. *Theory and Decision*, 52:171–199, 2002.

[96] E. Geigant and M. Stoll. Bifurcation analysis of an orientational aggregation model. *J. Math. Biol.*, 46:537–563, 2003.

[97] V. Gerasimenko. Towards kinetic equations of open systems of active soft matter. *arXiv:2412.09137v1 [math-ph] 12 Dec 2024*, 2024.

[98] V. Gerasimenko and Y. Fedchun. On kinetic models for the evolution of many-entity systems in mathematical biology. *J. Coupled Syst. Multiscale Dyn.*, pages 273–279, 2013.

[99] A. Gerisch and M.A.J. Chaplain. Mathematical modelling of cancer cell invasion of tissue: local and non–local models and the effect of adhesion. *J. Theoret. Biol.*, 250:684–704, 2008.

[100] R. Giniunaite, R.E. Baker, P.M. Kulesa, and P.K. Maini. Modelling collective cell migration: neural crest as a model paradigm. *J. Math. Biol.*, 80:481–504, 2020.

[101] R. Giniunaite, R. McLennan, M.C. McKinney, R.E. Baker, P.M. Kulesa, and P.K. Maini. An interdisciplinary approach to investigate collective cell migration in neural crest. *Dev. Dynam.*, 249:270–280, 2020.

[102] C.S. Gokhale and A. Traulsen. Evolutionary multiplayer games. *Dyn. Games Appl.*, 4:2270–2285, 2014.

[103] Q. Griette and S. Motsch. Kinetic equations and self–organized band formations. In *Active particles. Vol. 2. Advances in theory, models, and applications*, Model. Simul. Sci. Eng. Technol., pages 173–199. Birkhäuser/Springer, Cham, 2019.

[104] P. Grindrod. *The theory and applications of reaction–diffusion equations: Patterns and waves.* Oxford University Press, New York, 1996.

[105] H. Haken and J. Portugali. Information and self–organization: A unifying approach and applications. *Entropy*, 18:197, 2016.

[106] H. Haken and J. Portugali. Information and self–organization. *Entropy*, 19:18, 2017.

[107] E.T. Hall. *The Hidden Dimension.* Anchor Books Ed., 1966.

[108] J. Hofbauer and K. Sigmund. *The Theory of Evolution and Dynamical Systems.* Cambridge Univ. Press, 1988.

[109] J. Hofbauer and K. Sigmund. Evolutionary game dynamics. *Bull. Amer. Math. Society*, 40(4):479–519, 2003.

[110] E. Hollingdale, F.J. Pérez-Barbería, and D. McPetrie Walker. Inferring symmetric and asymmetric interactions between animals and groups from positional data. *PLos ONE*, 13(12), 2018.

[111] E.E. Holmes. Are diffusion too simple? a comparison with telegraph model of invasion. *The American Naturalist*, 142:779–795, 1993.

[112] T. Ihle. Kinetic theory of flocking: Derivation of hydrodynamic equations. *Phys. Rev. E*, 83:030901, 2011.

[113] G.J. Ikenberry, M. Mastanduno, and W.C. Wohlfarth. Introduction: Unipolarity, state behavior, and systematic consequences. *World Politics*, 61:1–27, 2009.

[114] E. Jäger and L. Segel. On the distribution of dominance in a population of interacting anonymous organisms. *SIAM J. Appl. Math.*, 52:1442–1468, 1992.

[115] K. Kang, B. Perthame, A. Stevens, and J.J.L. Velázquez. An integro–differential equation model for alignment and orientational aggregation. *J. Diff. Eqs.*, 246, 2009.

[116] S. Kaniel and M. Shinbrot. The Boltzmann equation. i. Uniqueness and local existence. *Commun. Math. Phys.*, 58:65–84, 1978.

[117] Ch.-Y. Kao and Y. Lou. Random dispersal vs. non–local dispersal. *Discrete Contin. Dyn. Syst. A,*, 26:551–596, 2010.

[118] D. Katsaounis, N. Harbour, T. Williams, M.A.J. Chaplain, and N. Sfakianakis. A genuinely hybrid, multiscale 3d cancer invasion and metastasis modelling framework. *Bull. Math. Biol.*, pages 1–41, 2024.

[119] W. Klingsch, C. Rogsch, A. Schadschneider, and M. Schreckenberg. *Pedestrian and Evacuation Dynamics.* Springer, 2008.

[120] M. Kolev, E. Kozłowska, and M. Lachowicz. A mathematical model for single cell cancer immune system dynamics. *Math. Comput. Model.*, 41:1083–1095, 2005.

[121] D. Kong, Y. Li, Z. Wang, and F.H. Sarkar. Cancer stem cells and epithelial–to–mesenchymal transition (emt) — phenotypic cells: Are they cousins or twins? *Cancers*, 3(1):716–729, 2011.

[122] S. Korenjak-Černe, N. Kejžar, and V. Batagelj. Clustering of population pyramids. *Informatica (Slovenia)*, 32:157–167, 2008.

[123] Y. Kozitsky. Stochastic branching at the edge: individual-based modeling of tumor cell proliferation. *J. Evol. Equ.*, 21(2):2081–2104, 2021.

[124] Y. Kozitsky and M. Röckner. A Markov process for a continuum infinite particle system with attraction. *Electron. J. Probab.*, 28:Paper No. 67, 59, 2023.

[125] Y. Kozitsky and A. Tanaś. Evolution of states of an infinite particle system with nonlocal branching. *J. Evol. Eq.*, 22(1):Paper No. 7, 25, 2022.

[126] T. Krueger, J. Szwabiński, and T. Weron. Conformity, anticonformity and polarization of opinion: Insights from a mathematical model of opinion dynamics. *Entropy*, 19:371, 2017.

[127] V.A. Kuznetsov, E. Heiko, and M.A.J. Chaplain. Editorial: Mathematical modeling and computational predictions in oncoimmunology. *Frontiers in Immunology*, 15, 2024.

[128] M. Lachowicz. Competition tumor–immune system. In *Proceedings of the VI National Conference Application of Mathematics to Biology and Medicine*, Zawoja, 2000.

[129] M. Lachowicz. *Multiscale Problems in the Life Sciences. From Microscopic to Macroscopic*, chapter Links Between Microscopic and Macroscopic Descriptions, pages 201–268. Springer, 2008.

[130] M. Lachowicz. Individually–based Markov processes modeling nonlinear systems in mathematical biology. *Nonlinear Anal. Real World Appl.*, 12(4):2396–2407, 2011.

[131] M. Lachowicz. Microscopic, mesoscopic and macroscopic descriptions of complex systems. *Prob. Engin. Mech.*, 26:54–60, 2011.

[132] M. Lachowicz. A class of microscopic individual models corresponding to the macroscopic logistic growth. *Math. Methods Appl. Sci.*, 2017.

[133] M. Lachowicz and M. Dębowski. Diauxic growth at the mesoscopic scale. *Entropy*, 22(11):1280, 2020.

[134] M. Lachowicz and H. Leszczyński. Modeling asymmetric interactions in economy. *Mathematics*, 8(4), 2020.

[135] M. Lachowicz, H. Leszczyński, M. Matusik, and Z. Szymańska. Integro–differential equations modelling the role of tendon cells in healing process. *unpublished work*, 2021.

[136] M. Lachowicz, H. Leszczyński, and M. Parisot. A simple kinetic equation of swarm formation: blow–up and global existence. *Appl. Math. Letters*, 57:104–107, 2016.

[137] M. Lachowicz, H. Leszczyński, and M. Parisot. Blow–up and global existence for a kinetic equation of swarm formation. *Math. Models Methods Appl. Sci.*, 27(6):1153–1175, 2017.

[138] M. Lachowicz, H. Leszczyński, and E. Puźniakowska-Gałuch. Diffusive and anti–diffusive behavior for kinetic models of opinion dynamics. *Symmetry*, 11(8), 2019.

[139] M. Lachowicz, H. Leszczyński, and K.A. Topolski. Self-organization with small range interactions: Equilibria and creation of bipolarity. *Appl. Math. Comput.*, 343:156–166, 2019.

[140] M. Lachowicz, H. Leszczyński, and K.A. Topolski. Approximations of kinetic equations of swarm formation: Convergence and exact solutions. *Appl. Math. Comput.*, 417, 126778, 2022.

[141] M. Lachowicz, M. Matusik, and K. Topolski. Population of entities with three individual states and asymmetric interactions. *Appl. Math. Computation*, 464, 128395, 2024.

[142] M. Lachowicz, M. Matusik, and K. Topolski. Asymptotic equivalence of nonlocal and local models of cell movement. *to appear*, 2025.

[143] M. Lachowicz, M. Matusik, and K. Topolski. Space–nonlocal description of cell movement. *Discrete Contin. Dyn. Syst. Ser. B.* 30(11): 4352-4362, 2025. doi:10.3934/dcdsb.2025058

[144] M. Lachowicz and T. Ryabukha. Equilibrium solutions for microscopic stochastic systems in population dynamics. *Math. Biosci. Engineering*, 10(3):777–786, 2013.

[145] M. Lachowicz and Z. Szymańska. Nonlocal models of biological phenomena. *Physics Life Rev.*, 12:76–77, 2015.

[146] M. Lachowicz and D. Wrzosek. Nonlocal bilinear equations. equilibrium solutions and diffusive limit. *Math. Models Methods Appl. Sci.*, 11:1375–1390, 2001.

[147] R. Landry and J.W. Moffat. Nonlocal quantum field theory and quantum entanglement. *Eur. Phys. J. Plus*, 139(71), 2024.

[148] A. Lasota and M.C. Mackey. *Chaos, Fractals, and Noise: Stochastic Aspects of Dynamics.* Springer, New York, 1994.

[149] A. Lasota and J.A. Yorke. Exact dynamical systems and the Frobenius–Perron operator. *Trans. Amer. Math. Society*, 273:375–384, 1982.

[150] C.T. Lee, M.F. Hoopes, J Diehl, W. Gilliland, G. Huxel, E.V. Leaver, K. McCann, J. Umbanhowwar, and A. Mogilner. Non–local concepts and models in biology. *J. Theoret. Biol.*, 210(2):201–219, 2001.

[151] H. Li, Y. Zhang, F. Yang, M. Zhang, H. Jing, Z. Li, A.I. Bature, and Q. Ma. Effect of spatial distribution of rubber powder on ductility of rubber asphalt from mesoscopic mechanical perspective: For bio-based materials recycling. *Ind. Crops Prod.*, 222:120103, 2024.

[152] J. Liu and Y. Wang. A quasilinear chemotaxis–haptotaxis model: the role of nonlinear diffusion and logistic source. *Math. Methods Appl. Sci.*, 40:2107–2121, 2017.

[153] X. Liu, W. Song, L. Fu, W. Lv, and Z. Fang. Typical features of pedestrian spatial distribution in the inflow process. *Physics Letters A*, 380:1526–1534, 2016.

[154] A. Lupica, P. Manfredi, V. Volpert, A. Palumbo, and A. d'Onofrio. Spatio–temporal games of voluntary vaccination in the absence of the infection: the interplay of local versus non-local information about vaccine adverse events. *Math. Biosci. Eng.*, 17(2):1090–1131, 2020.

[155] R. Mach and F. Schweitzer. *Multi–Agent Model of Biological Swarming*, volume 2801. Advances In Artificial Life. Lecture Notes in Computer Science, 2003.

[156] M.P. Machado Ramos, C. Ribeiro, and A.J. Soares. A kinetic model of t cell autoreactivity in autoimmune diseases. *J. Math. Biol.*, 79(6):2005–2011, 2019.

[157] H.A. Martînez-Berumen, G.C. Lopez-Torres, and L. Romo-Rojas. Developing a method to evaluate entropy in organizational systems. *Procedia Comput. Sci.*, 28:389–397, 2014.

[158] P. Mascheroni, C. Stigliano, M. Carfagna, D. P. Boso, L. Preziosi, P. Decuzzi, and B. A. Schrefler. Predicting the growth of glioblastoma multiforme spheroids using a multiphase porous media model. *Biomech. Model. Mechanobiol.*, 25:1215–1228, 2016.

[159] J. Monod. The growth of bacterial cultures. *Annu. Rev. Microbiol.*, 3:371–394, 1949.

[160] M. Morreau. Arrow's theorem. In Edward N. Zalta, editor, *The Stanford Encyclopedia of Philosophy*. Metaphysics Research Lab, Stanford University, Winter 2019 edition, 2019.

[161] W.A. Nelson, B. Joncour, D. Pak, and O.N. Bjørnstad. Asymmetric interactions and their consequences for vital rates and dynamics: the smaller tea tortrix as a model system. *Ecology*, 100(2), 2019.

[162] J.M. Oliveira, C. Ribeiro, A.J. Soares, and M.P. Machado Ramos. Equilibrium of a kinetic model of oscillating patterns for chronic autoimmune diseases. *Kinet. Relat. Models*, 2024.

[163] J.M. Oliveira and R. Travaglini. Reaction–diffusion systems derived from kinetic theory for multiple sclerosis. *Math. Models Methods Appl. Sci.*, 34(07):1279–1308, 2024.

[164] H.G. Othmer, S.R. Dunbar, and W. Alt. Models of dispersal in biological systems. *J. Math. Biol.*, 26:263–298, 1988.

[165] K.J. Painter. Mathematical models for chemotaxis and their applications in self–organisation phenomena. *J. Theoret. Biol.*, 481:162–182, 2019.

[166] K.J. Painter, T. Hillen, and J.R. Potts. Biological modeling with nonlocal advection–difuusion equations. *Math. Models Methods Appl. Sci.*, 34:57–107, 2024.

[167] S. Pal and R. Melnik. Nonlocal models in biology and life sciences: Sources, developments, and applications. *arXiv:2401.14651v1*, 2024.

[168] V.A. Panferov. On the interior boundary-value problem for the stationary Povzner equation with hard and soft interactions. *Ann. Scuola Normale Super. Pisa – Classe di Scienze*, 3, 2004.

[169] L. Pareschi and G. Toscani. *Interacting Multiagent Systems – Kinetic Equations and Monte Carlo Methods.* Oxford University Press, Oxford, 2013.

[170] M. Parisot and M. Lachowicz. A kinetic model for the formation of swarms with nonlinear interactions. *Kinet. Relat. Models*, 9(1):131–164, 2016.

[171] D.D. Patterson, A.C. Staver S.A. Levin, and J.D. Touboul. Spatial dynamics with heterogeneity. *SIAM Journal Appl. Math.*, 84(3):S225–S248, 2024.

[172] M. Perc, J. Gómez-Gardeñes, A. Szolnoki, L. M. Floría, and Y. Moreno. Evolutionary dynamics of group interactions on structured populations: a review. *J. R. Soc. Interface*, 10(80), 2013.

[173] B. Perthame. *Transport Equation in Biology.* Birkhäuser, 2007.

[174] B. Perthame. *Parabolic Equations in Biology. Growth, reaction, movement and diffusion.* Springer, Cham, 2015.

[175] B. Perthame. Kinetic equations and cell motion: an introduction. In *The dynamics of biological systems*, volume 4 of *Math. Planet Earth*, pages 239–263. Springer, Cham, 2019.

[176] M. Peyrard. Nonlinear dynamics and statistical physics of DNA. *Nonlinearity*, 17:R1–R40, 2004.

[177] M. Peyrard, S. Cuesta-López, and G. James. Modelling DNA at the mesoscale: a challenge for nonlinear science? *Nonlinearity*, 21:T91–T100, 2008.

[178] K. Pichór and R. Rudnicki. Continuous Markov semigroups and stability of transport equations. *J. Math. Analysis Appl.*, 249:668–685, 2000.

[179] S. Pillay, H.M. Byrne, and P.K. Maini. The impact of exclusion processes on angiogenesis models. *J. Math. Biol.*, 77:1721–1759, 2018.

[180] T. Płatkowski and R. Illner. Discrete velocity models of the boltzmann equation: A survey on the mathematical aspects of the theory. *SIAM Review*, 30(2):213–255, 1988.

[181] R.N. Rosecrance. Bipolarity, mutilpolarity and the future. *J. Conflict Resolution*, 10, 1966.

[182] M.D. Rosini. *Macroscopic models for vehicular flows and crowd dynamics: Theory and applications.* Springer Understanding Complex Systems, 2013.

[183] R. Rudnicki. Markov operators: Applications to diffusion processes and population dynamics. *Appl. Math. (Warsaw)*, 27(1):67–79, 2000.

[184] R. Rudnicki. *Models of population dynamics and genetics*, chapter From Genetics to Mathematics, pages 103–147. World Sci., New Jersey, 2009.

[185] R. Rudnicki. *Models and Methods in Mathematical Biology, In Polish.* IMPAN, Warszawa, 2014.

[186] R. Rudnicki and M. Tyran–Kamińska. *Piecewise Deterministic Processes in Biological Models.* Springer Briefs in Mathematical Methods, Cham, 2017.

[187] R.G. Rutledge and D. Stewart. Critical evaluation of methods used to determine amplification efficiency refutes the exponential character of real–time pcr. *BMC Molecular Biology*, 9:96–108, 2008.

[188] R.G. Rutledge and D. Stewart. A kinetic–based sigmoidal model for the polymerase chain reaction and its applications to high–capacity absolute quantitative real–time pcr. *BMC Molecular Biology*, 8:47–75, 2008.

[189] N. Sfakianakis and M.A.J. Chaplain. Mathematical modelling of cancer invasion: a review. In S. Takashi, C. Poignard, M.A.J. Chaplain, and V. Quaranta, editors, *Methods of Mathematical Oncology*, Springer Proceedings in Mathematics and Statistics, pages 153–172, Netherlands, 2021. Springer.

[190] A. Shafiee and F.T. Ghahramani. On the meaning of locality: the overlapping assumptions, quantum study. *Math. Found.*, 2:435–448, 2015.

[191] P. Sharma and N. Maffulli. Biology of tendon injury: healing, modeling and remodeling. *J. Musculoskelet Neuronal Interact*, 6(2):181–190, 2006.

[192] P. Siedlecki, J. Szwabiński, and T. Weron. The interplay between conformity and anticonformity and its polarizing effect on society. *J. Artificial Societ. Social Simul.*, 19(4):9, 2016.

[193] S.A. Silling and R.B. Lehoucq. Peridynamic theory of solid mechanics. In Hassan Aref and Erik van der Giessen, editors, *Advances in Applied Mechanics*, volume 44 of *Advances in Applied Mechanics*, pages 73–168. Elsevier, 2010.

[194] E.P. Solomon, L.R. Berg, and D.W. Martin. *Biology.* Belmont, CA: Brooks–Cole, 2011.

[195] R. Sommer. Studies in personal space. *Sociometry*, 22:247–260, 1959.

[196] F. Sommerlad and I. McLean (Eds.). *The political theory of Condorcet.* University of Oxford, Oxford, 1989.

[197] Q. Su, B. Allen, and J.B. Plotkin. Evolution of cooperation with asymmetric social interactions. *Proceedings Nat. Acad. Sci.*, 119(1), 2021.

[198] L.E. Szabó. The Einstein–Podolsky–Rosen argument and the Bell inequalities. *Internet Encyclopedia of Philosophy*, 2018.

[199] Z. Szymańska, M. Lachowicz, N. Sfakianakis, and M.A.J. Chaplain. Mathematical modelling of cancer invasion: Phenotypic transitioning provides insight into multifocal foci formation. *J. Computat. Sci.*, To appear, 2024.

[200] Z. Szymańska, C. Morales Rodrigo, M. Lachowicz, and M. Chaplain. Mathematical modelling of cancer invasion of tissue: The role and effect of nonlocal interactions. *Math. Models Methods Appl. Sci.*, 19(2):257–281, 2009.

[201] Y. Tao and M. Winkler. A chemotaxis-haptotaxis model: The roles of nonlinear diffusion and logistic source. *SIAM J. Math. ANal.*, 43:685–704, 2011.

[202] J.-P. Tassan and J.Z. Kubiak. *Asymmetric Cell Division in Development, Differentiation and Cancer, Series: Results and Problems in Cell Differentiation.* Springer Cham, 2017.

[203] J.P. Thiery, H. Acloque, R.Y. Huang, and M.A. Nieto. Epithelial–mesenchymal transitions in development and disease. *Cell*, 139(5):871–890, 2009.

[204] C.J. Thompson. *Mathematical Statistical Mechanics.* Princeton University Press, 1979.

[205] G. Toscani, C. Brugna, and S. Demichelis. Kinetic models for the trading of goods. *J. Statist. Phys.*, 151:1–18, 2012.

[206] S. Ukai and K. Asano. The Euler limit and initial layer of the nonlinear Boltzmann equation. *Hokkaido Math. J.*, 12:311–332, 1983.

[207] J.J. Vázquez. *The Porous Medium Equation. Mathematical Theory.* Oxford Sci. Publ., 2007.

[208] T. Vicsek, A. Czirók, E. Ben-Jacob, I. Cohen, and O. Shochet. Novel type of phase transition in a system of self–driven particles. *Phys. Rev. Lett.*, 75:1226–1229, 1995.

[209] M.C. Vinerean. *Discrete Kinetic Models and Conservation Laws.* PhD thesis, Karlstad University, Sweden, 2005.

[210] V. Volpert, M. Banerjee, A. d'Onofrio, T. Lipniacki, S. Petrovskii, and V.Ch. Tran. Coronavirus – scientific insights and societal aspects. *Math. Model. Nat. Phenom.*, 15:E2, 2020.

[211] R.H. Wagner. What was bipolarity? *International Organization*, 47:77–106, 1993.

[212] L. Wang, Ch. Mu, and X. Hu. Boundedness in a quasilinear chemotaxis–haptotaxis system with logistic source. *Math. Methods Appl. Sci.*, 40:3000–3016, 2017.

[213] Z. Wang, C.T. Bauch, S. Bhattacharyya, A. d'Onofrio, P. Manfredi, M. Perc, N. Perra, M. Salathé, and D. Zhao. Statistical physics of vaccination. *Physics Reports*, 664:1–113, 2016.

[214] J.R. Weeks. *Population: An Introduction to Concepts and Issues.* Wadsworth Publishing, 2015.

[215] A.D. Wentzell. *A course in the theory of stochastic processes.* McGraw-Hill International, 1981.

[216] P.G. Wright. Entropy and disorder. *Contemporary Phys.*, 11:581–588, 1970.

[217] Z. Wu, T. Yin, H. Li, and J. Zhao. *Nonlinear Diffusion Equations.* World Sci., 2001.

[218] www url. Lab manual: Measuring DNA melting curves. `http://scripts.mit.edu/~20.309/wiki/index.php?`, 2013.

[219] G. Yang, B.B. Rothrauff, and R.S. Tuan. Tendon and ligament regeneration and repair: Clinical relevance and developmental paradigm. *Birth Defects Res C, Embryo Today*, 99(3):203–222, 2013.

[220] J. Yang, P. Antin, G. Berx, C. Blanpain, T. Brabletz, M. Bronner, K. Campbell, A. Cano, J. Casanova, G. Christofori, Sh. Dedhar, R. Derynck, H.L. Ford, J.G. Fuxe, A. García de Herreros, G.J. Goodall, A.K. Hadjantonakis, R.Y. Huang, Ch. Kalcheim, R. Kalluri, Y. Kang, Y. Khew–Goodall, H. Levine, J. Liu, G.D. Longmore, S.A. Mani, J. Massagué, R. Mayor, D. McClay, K.E. Mostov, D.F. Newgreen, M.A. Nieto, A. Puisieux, R. Runyan, P. Savagner, B. Stanger, M.P. Stemmler, Y. Takahashi, M Takeichi, E. Theveneau, J.P. Thiery, E.W. Thompson, R.A. Weinberg, E.D. Williams, J. Xing, B.P. Zhou, and G. Sheng. Guidelines and definitions for research on epithelial–mesenchymal transition. *Nature Rev. Molecular Cell Biology*, 21(6):341–352, 2020.

[221] H. P. Young. Condorcet's theory of voting. *American Political Sci. Review*, 82(4):1231–1244, 1988.

Index

FSC
www.fsc.org
MIX
Papier | Fördert gute Waldnutzung
FSC® C083411

Kolibri 360 GmbH
Ferdinand-Jühlke-Straße 7
99095 Erfurt, Deutschland
produktsicherheit@kolibri360.de